SELECTED POEMS

SELECTED POEMS

1956–1994

Chris Wallace-Crabbe

Oxford   New York   Melbourne
OXFORD UNIVERSITY PRESS
1995

Oxford University Press, Walton Street, Oxford OX2 6DP
Oxford New York
Athens Auckland Bangkok Bombay
Calcutta Cape Town Dar es Salaam Delhi
Florence Hong Kong Istanbul Karachi
Kuala Lumpur Madras Madrid Melbourne
Mexico City Nairobi Paris Singapore
Taipei Tokyo Toronto
and associated companies in
Berlin Ibadan

Oxford is a trade mark of Oxford University Press

Selected Poems 1956–1995 first published in Oxford Poets
as an Oxford University Press paperback 1995

British Library Cataloguing-in-Publication Data
Data available

Library of Congress Cataloging-in-Publication Data
Wallace-Crabbe, Chris.
[Poems. Selections]
Selected poems, 1956–1995/Chris Wallace-Crabbe.
p. cm.—(Oxford poets)
I. Title.
PR9619.3.W28A6 1995 821'—dc20 94–26339

ISBN 0-19-282412-0

1 3 5 7 9 10 8 6 4 2

Typeset by Rowland Phototypesetting Ltd
Printed in Hong Kong

For Peter Steele

'Bless what there is for being.'
—W. H. Auden

ACKNOWLEDGEMENTS

It is not easy deciding what to put into a book which represents your past life. In making this selection, I have been helped by the judgements of friends and anthologists, as well as deciding for myself which poems still 'work'; that is to say, dance off the page. Some attention has also been paid to asking the question, which poems most clearly bear witness to past selves. My warm thanks must go to Lisa O'Connell, who helped to prepare the manuscript, and to Jacqueline Simms, at Oxford and Greenwich, for her continued support as well as for her good advice.

From *The Music of Division* (1959) until *The Emotions are not Skilled Workers* (1980), including an earlier *Selected Poems* (1973), my poems were published by Angus & Robertson; since 1985, by Oxford University Press, in Oxford, New York, and Melbourne.

For permission to reprint new poems, my thanks are due to the editors of *The Age*, *Agenda*, *Arena*, *Harvard Review*, and *Voices*.

C.W–C

CONTENTS

From *The Music of Division* (1959)
  Citizen 1
  Ancient Historian 2
  Practical Politics 3
  The Wife's Story 4
From *In Light and Darkness* (1963)
  A Wintry Manifesto 5
  In Light and Darkness 7
  Every Night about this Time 9
  The Swing 10
  Wind and Change 10
  Melbourne 11
  A Sentimental Education 12
From *The Rebel General* (1967)
  Carnations 13
  Nature, Language, the Sea: an Essay 13
  Traditions, Voyages 15
  The Apparition 16
  The Threat 17
From *Where the Wind Came* (1971)
  The Centaur Within 18
  The Joker 19
  In Hay Fever Time 20
  Losses and Recoveries 21
  Evening at San Miguel 23
  Other People 24
From *The Foundations of Joy* (1976)
  Last Page from an Explorer's Journal 25
  Real Estate 26
  The Men of Meat 27
  The Wild Colonial Puzzler 28
  It 29
  Again 29
From *Selected Poems* (1973)
  Meditation with Memories 30
  Hey There! 31
  The Collective Invention 32
  Keeping One's Head 33

From *The Emotions Are Not Skilled Workers* (1980)
    Genesis 34
    The Shape-Changer 36
    Bennelong 37
    New Carpentry 38
    Puck in January 39
    Pantuns 40
    In the Wilderness 41
    Old Men during a Fall of Government 42
    Now that April's Here 43
    The Mental Traveller's Landfall 44
    Introspection 45
From *The Amorous Cannibal* (1985)
    Gaspard de la Nuit 47
    Mind 48
    Abhorring a Vacuum 49
    Nub 50
    Exit the Players 51
    The Bits and Pieces 52
    Panoptics 63
    Practitioners of Silence 64
    Sacred Ridges above Diamond Creek 65
    A Stone Age Decadent 66
    The Amorous Cannibal 69
From *I'm Deadly Serious* (1988)
    The Well-Dreamed Man 70
    There 71
    The Starlight Express 72
    A Glimpse of Shere Khan 73
    Genius Loci 74
    Sonnets to the Left 75
    The Mirror Stage 80
    Stardust 81
    God 83
    The Lorelei 84
    The Thing Itself 85
From *For Crying Out Loud* (1990)
    They 86
    An Elegy 87
    The Life of Ideas 88
    The Inheritance 91
    Mental Events 92

River Run 93
The Evolution of Tears 94
The Sound 95
For Crying Out Loud 96
Puck Disembarks 97
The Bush 98
And the World Was Calm 99
From *Rungs of Time* (1993)
Like Vibrations of a Bell 101
Trace Elements 102
Afternoon in the Central Nervous System 103
Opera with Phantoms 104
Free Will 105
Call Me Jack 106
Garth McLeod Brooding 107
Sunset Sky near Coober Pedy 109
Looking down on Cambodia 110
Reality 111
Dusky Tracts 112
New poems
Ode to Morpheus 113
Morpheus Replies 114
What Are These Coming to the Sacrifice? 115
Why Do We Exist? 116

CITIZEN

The roofs of cars were crusted thick with frost
And ice was notable on every pool;
Blue children stalked in overcoats and gloves
Past the Immaculate Heart of Mary School.

And where he walked the traffic rolled one way
Toward the solemn ritual of work—
While hasty breakfasts rumbled in those cars
Whose flying mud had stopped him with a jerk.

On villa, cottage, sweet-shop, bungalow,
Schoolroom and Gentlemen's, the morning grinned
Facetiously. He numbly picked his way
Among the houses where the world had sinned.

Now schizoid barbers packed their shaving gear,
Uncomfortable lovers caught the train
And everyone rushed. It was at times like this
That decent violence beset his brain.

'Complacent city with your brazen bells
And morning song . . .' He called for words to cease,
For citizens to know their proper hells
And anger to bloom green upon the trees.

ANCIENT HISTORIAN

The drowsy Herodotus holds him there;
He drums his knuckly fingers on the page
And glares around the library again
With indignation that is almost rage.
A bookish valour arms him to the teeth;
His abstract love of war and copulation
Is turbine to his blood and, spurning print,
He seethes with all the fire of one ambition:

To lead a narrow band of dead Italians
In through the door past Dr Johnson's bust—
Stout Marius, and Catilina soiling
His well-bred feet in the rebellious dust,
And Clodius encouraging his mob
To yell out *'Panem et circenses!'* here.
Perhaps that might disturb a few thick heads
And blow some dust out of the atmosphere.

Vitality, he muses, is the stair
By which we climb to join the armoured great,
The toll exacted by our family gods,
The virtue proper to both man and state.
But, caught in sudden vision, he recalls
A town once founded on vitality,
Rome, sprawling and gone rotten at the core,
And the lean Goths encroaching silently.

PRACTICAL POLITICS

Suppose this edict were proclaimed today—
Let poets and philosophers be kings.

Let poets and philosophers be kings
Four days a week while under careful guard

By senior public servants and the press;
Suppose this edict were proclaimed today,

Four days a week while under careful guard
Sages could walk the parliamentary round

And make some tiny changes in the state
Until they had been sobered by such things.

Suppose this edict were proclaimed today
By senior public servants and the press—

Let poets and philosophers be kings
Four days a week while under careful guard—
Sages could walk the parliamentary round
Until they had been sobered by such things.

THE WIFE'S STORY

What with his taking small bets on the footy,
While keeping out a sharp eye for the Bulls,
And what with laying off among other bookies
Who masqueraded as both knaves and fools;

What with his ringing up those drunken mates—
The one who sang 'Jack Doolan' in the bar,
The fat one and the one with smelly feet—
And talking to the bloke who tuned his car;

What with his ringing up the spirit merchant
And with those calls he made too low to hear
But which, she long suspected, were his urgent
Demands on girls with easy yellow hair;

And what with the invitations he'd refuse,
The telephone was never hers to use.

A WINTRY MANIFESTO

It was the death of Satan first of all,
The knowledge that earth holds though kingdoms fall,
    Inured us to a stoic resignation,
    To making the most of a shrunken neighbourhood;

And what we drew on was not gold or fire,
No cross, not cloven hoof about the pyre,
    But painful, plain, contracted observations:
    The gesture of a hand, dip of a bough

Or seven stubborn words drawn close together
As a hewn charm against the shifting weather.
    Our singing was intolerably sober
    Mistrusting every trill of artifice.

Whatever danced on needle-points, we knew
That we had forged the world we stumbled through
    And, if a stripped wind howled through sighing alleys,
    Built our own refuge in a flush of pride

Knowing that all our gifts were for construction—
Timber to timber groined in every section—
    And knowing, too, purged of the sense of evil,
    These were the walls our folly would destroy.

We dreamed, woke, doubted, wept for fading stars
And then projected brave new avatars,
    Triumphs of reason. Yet a whole dimension
    Had vanished from the chambers of the mind,

And paramount among the victims fled,
Shrunken and pale, the grim king of the dead;
    Withdrawn to caverns safely beyond our sounding
    He waits as a Pretender for his call,

Which those who crave him can no longer give.
Men are the arbiters of how they live,
    And, stooped by millstones of authority,
    They welcome tyrants in with open arms.

Now in the shadows of unfriendly trees
We number leaves, discern faint similes
      And learn to praise whatever is imperfect
      As the pure breeding-ground for honesty,

Finding our heroism in rejection
Of bland Utopias and of thieves' affection:
      Our greatest joy to mark an outline truly
      And know the piece of earth on which we stand.

IN LIGHT AND DARKNESS

To the noonday eye, light seems an ethical agent,
Straight from the shoulder, predictable, terribly quick, though
It climbs in a curve to space's unlikely limit,
   Thus posing a problem for mathematicians and God,
Whatever He turns out to be and wherever His dwelling.
Rich is the clotting of gold in the late afternoon or
Turning to twilight, one last blaze of watery colour
   Where man can project his false dreams in figures of light,
Pretending that all his environment loyally loves him,
Seas of ice and burning plains.
        Too easy
Bending nature before us, but light is defiant
   Coming by night from dead stars with terrible speed.
So is our planet rebuked, and we meet in mirrors
Desperate masks, eyes of imprisoned strangers
And lips that open to say, 'We are only mirages,'
   While the lawns outside are green and the roses real.

You dub the sun a realist? Then it will plague you,
Wading with stork-legs into the green water,
Lending the oak-bole moss its leopard shadows
   Or streaking paddocks darkly with sunset sheep
Fifty feet long.
        Nothing is quite so rococo
As dawnlight caught on a fishscale formation of cirrus,
Nor quite so romantic as one gilded westering biplane.
   We just don't live in a hard intellectual glare.

No one, of course, endures darkness or daylight entirely:
Cyclic change betrays our terrestrial journey
And what looked like trees at dawn turn out to be crosses,
   Suddenly black as the florid sun goes down,
And maybe at midnight resemble gallows or statues
As we slip past, childlike, alone.
       No wonder
Sly poets admiringly use, but will not warm to
   The subtle machines that teach our world to spin.

These are too steady for reverence: gods must be changeable
Since worship is saved for what we cannot govern;
Gods push up grass, channel the dense sap through pine-boughs
          And amass the pewter cloudbanks of summer storm.

Once I awoke, a child in a chill mountain morning,
To see the small town—undreamed transfiguration—
Mantled in white, its slate and stone and timber
          Bright with that foreign cloak of innocence.
And I walked into old-world beauty; the gilt sun rising
Fell in a garden where time itself was congealing
As we shaped wonderful igloos under the stringybarks
          And took no account of thaw. But noon came on.
Something was lost in the brown receding slush there
Which has not returned, something other than childhood:
A notion, rather, of clear crystalline standards
          Freezing life to one shape, like a photograph.

Yet this is the point:
                              a photograph leaves out living
For Eton crops, old blazers, baptismal lace or
Some late-Victorian smile turned stiff and waxy,
          All arabesques, but never the heart of the thing,
Which is neither good nor bad, but one maze of motion
Through which we dance, into and out of the darkness
To tireless music: motes in the curled winds' breathing
          And more than motes, faced with the corners of choice.
And so at night below all the brilliant clusters
Of lamps in the sky, the living, dead and dying
Poised in their dance, I cling to the crust here stolidly
          And pray for a perfect day.
                              Out in the cold
Of hoarfrost and starlight, we fear for tomorrow's choosing
And cherish dreams made in meticulous patterns;
But come tomorrow, we will neither be Christ nor Gandhi
          But will breathe this polluted air and rejoice with the birds.

All that I ask is that myriad lights, ever changing,
Continue to play on this great rind of ranges and valleys,
Flooding the vision of dreaming dwellers in cities
          Who walk out in summer pursuing something to praise.
          We will neither be simple nor clear till the end of our days.

8

EVERY NIGHT ABOUT THIS TIME

Smoke in a fine plume curls against the blind,
One thin cigar crumbling to ash;
Out of the night a rainy wind
Batters and soughs for love against the sash.

Now sorcerers and shamans walk abroad,
Tear up the lawns, sever the wires,
Put leading punters to the sword
And shake our bluestone churches by their spires.

Yet poets squat alone by reading-lamps,
Tugging upon uneven hair
And, chewing at their pencil stumps,
Madly seek rhymes for incest and despair.

Such words as love, cried Auden bitterly,
Have in our time been soured, debased,
And knew, for all his flippancy,
The pits and dangers that a world has faced.

Though history plants the villas in their shrubs,
Bulking eternally apart,
Late lights linger in the backs of pubs
And couples in dark Fords lay bare the heart.

THE SWING

On a swing at midnight in the black park. Between poplars
which are towers of light for a hidden street lamp and inky
she-oaks my arc is maintained. From lighter to darker I go,
from dark to light; but only, as ever, to return.

Here we live in the imperfect syntax of light and darkness;
wanting to write a sentence as perfect as the letter o in praise of
things. For things exist supremely; all our values cohere in
things.

The austere prose which could outline the world with a
physicist's clarity never arrives. We move through the fugal
elaboration of leaves, through centuries of drowning flowers.
Unsatisfied, uncertain, I am swinging again tonight in the park.

WIND AND CHANGE

It is the body swaying on its stalk,
The living bloom aware of light,
Even those hands in motion as in leaf
That shake me so
Who draw near in disembodiment, delight
Even: crossing the furry lawn,
Butting through wind, impelled by some belief
In the dazzling rays of a world made fresh.
Song, even tree,
Are imperfect analogies
For green assertion in the wind's teeth,
For all this warmth in the day's eye,
For everything disordered, all in leaf,
Laughing and blowing.

MELBOURNE

Not on the ocean, on a muted bay
Where the broad rays drift slowly over mud
And flathead loll on sand, a city bloats
Between the plains of water and of loam.
If surf beats, it is faint and far away;
If slogans blow around, we stay at home.

And, like the bay, our blood flows easily,
Not warm, not cold (in all things moderate),
Following our familiar tides. Elsewhere
Victims are bleeding, sun is beating down
On patriot, guerrilla, refugee.
We see the newsreels when we dine in town.

Ideas are grown in other gardens while
This chocolate soil throws up its harvest of
Imported and deciduous platitudes,
None of them flowering boldly or for long;
And we, the gardeners, securely smile
Humming a bar or two of rusty song.

Old tunes are good enough if sing we must;
Old images, re-vamped *ad nauseam*,
Will sate the burgher's eye and keep him quiet
As the great wheels run on. And should he seek
Variety, there's wind, there's heat, there's frost
To feed his conversation all the week.

Highway by highway, the remorseless cars
Strangle the city, put it out of pain,
Its limbs still kicking feebly on the hills.
Nobody cares. The artists sail at dawn
For brisker ports, or rot in public bars.
Though much has died here, little has been born.

A SENTIMENTAL EDUCATION

Prismatic crystals, red unshrouded coals
Glowing and tumbling in the grate,
An atlas where the file of Andes snaked
Through pocket states in tan and blue and green,
The click of dense equations coming out
As naked logic ended with a bow
And proved his faith—a world was built on these:

The Western Indies touched with rum and gold
Crisply inscribed upon a printer's sea,
The Carolinas rocked by Edward Teach,
Statistics, batting averages and graphs
That soared like Himalayas on a page
Meshed with green lines—jagged precision there!—
And then that most elaborate wonderland,
The periodic table on a chart
Where every element observed his place,
His weight, his quality; those ninety-two
Were allegorical protagonists;
Zinc, argon, sulphur, calcium and gold
Were knights for whom the week itself would ride;
Flowers of wild colour in the Bunsen flame,
The blank facility of *qu'est-ce que c'est*,
Mark Twain, Keith Miller, Alexander Pope,
Dichromate crystals like unshrouded coals,
Gleaming and growing in the light
Week after week . . . .

CARNATIONS

From the green cartridge an explosion of
Substantial pink;
               and energies made clear
That drive our crinkled flesh into this world,
Unfurl the layers of a life
And thrust with ragged edges to the light.

NATURE, LANGUAGE, THE SEA:
AN ESSAY

*But what of the artist? Has he either knowledge
or correct belief?*—The Republic

With such worn currency of models as
               Rivers and steps and roads, the mind
Does much of its diurnal labour;

Analogies its glory and its task,
               Into the shambling mess we call
Creation it plunges, asking 'Why?'

For the world is wonder, is profusion,
               A boundless brilliant orchard of
Sun-licked, thunder-shaken strangeness,

But nobody can claim it makes good sense
               Or testifies to providence:
O, the mistakes of a creator!

Nature, language, sea: our great examples
               Of what must always be rebuked
By the modest radiance of art.

Where the thick ridges of a mountain range,
               Ruffled with unbroken forest
And innocent of bridge or roadway,

Roll like a stiff surf out from under you
       In one great Burkean rhythm
To shout, Sublime, Sublime, O Sublime,

And the gullies brim with a steely blue mist,
       Is there, as your heart takes pause,
Anything heard but a hymn of chaos?

There are no sermons in the Great Divide;
       The boundless overflows and
Its tide is no river of symbols;

Yet symbols must rise glittering from some
       Quick river or sullen current,
Dying as soon as their surface dries,

And sound roots must aspire from common soil
       Earning a sudden right to bear
The fire-featured apples of the sun.

Nature, language, the sea; and opposite
       These fluent fields of energy
We feel that we are bound to chisel

The small hard statues of our poetry
       Which bear, like a sheen on marble,
Burnished assurance of being right,

All the assurance that we come to know
       Under the name of Form, and know
In the strong humming of completeness

Wherein the formal is at last the good.
       Both passionate and moral is
Music which has passed into a shape,

A river which has found its dimpling course
       Through the golden fields of Chronos,
Sounding not of loss but clarity.

TRADITIONS, VOYAGES

(Australian to Jamaican)

Your barriers confront me from the page,
History, news, black hubbub from afar,
Your spice-wind echoes breathing upon our shores
    But all so faint, so faint.

We had, perhaps, our Middle Passage too,
Irishman and felon stank in the holds
When frail barques ferried the Enlightenment,
    Already dim and faint.

In darker parts of the map a growing swell
May overturn the lumbering Ships of State;
Diminished ripples bend the seaweed where
I walk the shore,
              hear a gull cry,
    Discordant, fretful, faint.
My heel grinds in the white sand
As I am driven to confront
Drab skyline, yellowing papers, a fat land.

THE APPARITION

After the entertainments of the night
somewhere in limbo he had lost his way,
a dark suit in a darker street
swinging toward a nameless day.

Leaves ran murmuring from his feet
and agonies of perfume round his head
drifted from shrubs he could no longer name,
from petal, stem and calyx of the dead.

A Norfolk pine sprang skyward in front of him;
he knew it well, had climbed it as a child,
so when he saw somebody quit its port
of shade he was already reconciled

to the small bent figure with collar caught
beneath her chin by a massive cameo,
to glasses crooked on their bridge,
to the known voice, its unperturbed hello.

'Michael,' she said. Light as a hovering midge,
she moved beside him through bluish air.
'Those cannas which you planted are in bloom,
the daisies, too; but weeds are everywhere.

'For seven months I wilted in a room.
It was your name that rose upon my breath,
and yet my favourite grandson never came
into the chamber of my death.'

As greasy starlings cried for shame
and a twig cracked where his footstep fell
Michael addressed the rotting universe:
'I was only a child. How could I tell?'

Her ghost had vanished. Deep grey like a hearse,
the sky bore downward with its close-webbed net
and every leaf was echoing,
'You always found it easy to forget.'

THE THREAT

Coals glow; flames flicker on the backs of books;
Giovanni shifts his head
And squints a little, trying to isolate
So faint a note of dread:
Not the metallic pip of wheeling bats,
Not a wind-scuffled bough
And not the bite of wheel in gravel road-edge.
Who made the sound? Where? How?
What moves beyond his sight?

Good books, wry memoirs, will not answer this
Occasion of pure fright,
For who can tell what casual hour will bring
Out of the bruise-blue night
A marble parent thundering at the door

THE CENTAUR WITHIN

Your tight nets dredge the water
for leaves,
for broken moons,
but this you will not catch.

Moralities close in,
wide shades
fill the garden
drowning from wall to wall,

but I frisk on my track,
centaur,
and my hard heels
clatter on the cobbles

blocks away already,
quick, free.
Although complete
your picture leaves me out;

gather all to granite
greyness,
a proper scheme,
and still you miss this flame

which burns for nobody
at all
but its cruel self,
flaring *sum ergo sum*:

I pay no tax or trust,
no friend
sustains me here;
I am essential self.

THE JOKER

Self
is the springer,
the limber light evader shooting through
past all our fences;

look, see how
this fellow gets through the day
scuffling with his heavy neighbour,
personality,

who yet again
sits down here glum and firm
(paid agent of stability),
a hulking lout

waiting
to get his big hands
on silly self at last:
half his luck!

Dear self,
you flip out of anyone's grip
like a wet watermelon seed . . .
there you are . . .
                              off again.

Chevalier,
I delight in your quick spring heels,
and yet baffled in the end
I fall back

unable
to keep with you, catch you,
stick to the smashing pace you have set.
You tear me apart.

IN HAY FEVER TIME

Spring breaks down
the dikes of years
letting us into ourselves,
letting our dreams
flap out strongly
in a real wind.

A sickly photinia
gathers in moths by night
while the air is burdened
with palpable temptations:
mown grass, wistaria,
pittosporum, verbena.

Every seven years
we grow a new body,
an ark for the selfish flame;
we open up high territories,
come on strange flavours,
find new enemies to hurt.

Scents pass,
that much is sure at least,
but who sets a limit to
flowers, friendships, verities,
all running away down
at different speeds to death?

The grass-heads are thickening.
Subject to sinus and pollen,
beleaguered by flesh,
a would-be immortal spirit
burns fiercely away
now in the teeth of darkness.

LOSSES AND RECOVERIES

I

There he goes, went, catch him,
small boy in a beret, nervously smiling, taken
for tourist walks on the wrong end of a leash
in that snow-white stretch of life I can't remember:
half-timbered Munich, strudelplatz Berlin,
droppings of history mounded high above me,
over those pigeons the Leica had me feed.
Where's it all gone, that Deutschland culture-fodder
that I soaked up in travel, a willing sponge?

Crusty surfaces in half-tone,
brick, stone, snow, inked crosses marking
a bit of myself on this or that hotel
and here's the window where I heard the Rhineland
cobbles re-echo to military geese,
but I can't know whatever I could tell,
what strands were plaited by which war, which peace.

II

A drive flashes from the thick meat of my bat,
leaving extra cover for dead. All around us,
touched by a pastoral brush,
midges hover and glint:
chiaroscuro daubs the river-gums.
The serene hour brims with oceanic feeling,
drowning these four green ovals deep as dream
in which I move at ease, for Morrison's off-spinners
aren't going to turn an inch this afternoon.

Goings and dwindlings: my stupid adolescence,
bone-dry years of hollowness and blank,
thirsting, fills out again with fields and games
till my lifelong model
of happiness or poise becomes
a well-timed leg glance taken off my toes.

III *For Bruce Dawe*

Film has no tenses, the latest pundit says,
poems have tenses and nostalgias though
like anything, and when I get to think
of the mid-fifties, flashing through my slides,
I see you slope past Chemistry, blue-chinned
in military shirt and a maroon
figured art-silk tie with look-I'm-right
Ciardi's *Inferno* splitting away to cantos
and well-thumbed pages in your jacket pocket,
verse on your tongue of mice in evening dress.

In memory's yellow eye it's always summer,
nobody ever worked, the grass grew thick,
coffee cups, unlike women, had no bottoms
and there you grin, old-fashioned Carlton shepherd
dry-wittedly enjoying
the Arcadian lull between Joey Cassidy stories.

IV

Looms up the liner's rivetty white side,
streamers droop overhead,
our good friends go. There are many folk we love
this way and that, struggling to find a balance
where the high side fills with a charge of shadows
and wharfside waters lap quietly below.
Dulled along splintery boards we walk away.

Melbourne holds us: hands, lips, bodies, all
that we are it feeds—and feeds upon;
many would go, but drag the city with them
world-wide, wherever they push and flee.
Here, look now, all this is ourselves and
up over the dull blue skyline dangles
a bright eastering T-jet the colour of hope.

EVENING AT SAN MIGUEL

Suddenly, at shadow, the plaza fills with birds,
Fills to the very brim,
To the eaves of surrounding mansions
With a draught of turbulent garrulous birds
Disturbing clipped laurels and vintage air.

Suddenly, with late sun, the plaza is full of people
Pouring out of their houses,
Uttered from small streets,
And girls go this way and the boys go that way
Around the laurels with an easy pace.

Late, now, a sunlight emanates from stonework,
Radiates back to the sky
As my body, foreign, relaxed and eager,
Gives back experience to the walls around it,
Re-lights the laurels, unbreathes a stone-warmed air.

OTHER PEOPLE

In the First World War they . . .
Who were *they*? Who cares anymore? . . .
Killed four of my uncles,
So I discovered one day.

There were only four on that side of the family
And all swept away in a few bad years
In a war the historians tell us now
Was fought over nothing at all.

Four uncles, as one might say
A dozen apples or seven tons of dirt,
Swept away by the luck of history,
Closed off. Full stop.

Four is a lot for uncles,
A lot for lives, I should say.
Their chalk was wiped clean off the slate,
The War meant nothing at all.

War needs a lot of uncles,
And husbands, and brothers, and so on:
Someone must *want* to kill them,
Somebody needs them dead.

Who is it, I wonder. Me?
Or is it you there, reading away,
Or a chap with a small-arms factory?
Or is it only *they*?

LAST PAGE FROM AN EXPLORER'S JOURNAL

Approaching from the south we lost Jacoby.
Alluvial soils gave out. The fevers went.
We climbed through furze towards the low Bicuspids.
It was then, as I recall, that Dr Spade
had trouble with the ponies and their harness.
Here and there anthracite came through the shale.
The whole thing is an ancient rift valley.
It was very dull.
                 Tempers grew worse
until, that Wednesday, our track led round and up
to the east face, and we were in the Canines,
which, with a sudden turnabout of weather,
gleamed high and jagged. Clouds brighter than snow
scarved them here and there. We ate small berries,
gathered campanula and woundwort. Murphy died of boils.

REAL ESTATE

The structure
of the future
has no pilaster,

vault or architrave.
Its prospects always strive
with what we have,

hammering our hopes
into queer shapes
and breakneck slopes.

Can lives afford
the perfunctory, sad,
galvo-and-weatherboard

sheds and halls
on the wrong hills
when all investment fails,

listening to the broadcast
fall of wickets
with tea and biscuits?

Listen, mister,
the nightly planner
of your future
is death's well-known brother.

THE MEN OF MEAT

'Butchers are all so healthy and nice looking,
What beautiful pink skins they have,'
I said one Saturday, one summer,
In the main street of a Gippsland town.

Full of protein, briskly smiling,
In their laundered linen whites
They stood brandishing their cleavers
And well-honed knives.

A wire screen kept the flies at bay,
Customers kicked sawdust round
While making crucial weekend choices,
Chops, T-bone, a roast;

While the boss marked incisions
In a forequarter, neat as pie,
Or else flopped steaks on bits of paper
('Thank you, madam. Will that be all?')

And his logical neat window
Like a flower garden, red and white
Varied with the odd black pudding,
Really turned me on.

'I'd *almost* like to be a butcher'
Was the thought ran through my brain
But my senses were already loping
Ahead to the lunchtime barbecue.

THE WILD COLONIAL PUZZLER

Padding through the grounds of a great house,
meaning to make sense
of all this landed equanimity,
lawn, gazebos, coaltits, wagtails
and garlicky water, I want to tell you
beware of Australians bearing gifts.
I bring along the stink of restlessness
like an infection, like a second skin;
but to say this, friends, would be to patronize,
assuming you don't know what you know.

Hopkins lived here.
I can't pretend to equal his unrest
nor his quick marvellous inwardness
with spray, pod, quoin, penumbra, ripple,
everything unresting in creation,
but cross the garden, quiet as a fieldmouse
to that waterlily-splintered pond
where a bent classical figure, Prometheus maybe,
looks up to make sure
promised Lancastrian thunderstorms haven't broken.

IT

I am an eye,
sheer mechanical eye.
Although machine, I show you a world
the way only I can see it,
and ever just a little grainy.

I am released from starchy humanity
being in constant movement
whoosher and zoomer
approaching and pulling away from objects:
I creep underneath them.
At Moonee Valley races I gallop
beside a rapid horse's mouth.
I fall and rise
with the falling and rising bodies
devoid of libido. I am a god
or an eye
prepared to inherit the earth.
Blandly I sail to the Alimentary Canal
and blink at Phnom Penh,
but cannot taste a wine or smell these flowers.

AGAIN

March is coming;
midnight returns.
The choir of angels
packs up and feathers home
and with the faintest of sighs
earth rolls over
in her beautiful sleep;
one breast is flattened against dry stubble
and one dark hip
thrust upward among the stars.

MEDITATION WITH MEMORIES

All over the world things are going on at once,
Space–time collapses, the caravels inch west,
I am a boy again reading strange headlines,
ELEPHANTS TRIUMPH: HANNIBAL CROSSES ALPS,
The alphabet is invented, the printing press,
I move to another school,
Doomed soldiers run through dust toward the Turks
And Billy Hughes is ravaging Versailles.

Evening star,
Hungry Vesper,
You punctured a gloaming sky,
Steady against the motion
Of our dark southwarding carriages,
Solemnly, brilliantly, there
At my right hand.

What is our memory that it deposits
In such uncobbled alleys of the mind
In charges or solutions fond events,
Honey against the winter coming on?
Random and miniature its family-album,
Brilliant the slides but dim the prophecy
That schema of lost action have to offer:
Memory hugs the heat against itself.

Evening star,
You swayed along beside us
Mile after mile
Along the dry sierras
And westward villages foundering into night,
Clean profile,
Sparse terrain,
The miles that travellers devour
In a gust of physical hunger
To quest and to arrive
Anxiously, ardently, soon,
In rough towns waiting for new selves,
Waiting to shape our ways,
The outlines of a future we prepare.

HEY THERE!

On Ferntree Gully line
the usual motivations
are shunting about:

hard ownership as
a style of style,
anxiety's children

driven up the creek,
avuncular ghosts
or the blaming habit.

Who can beat down
a teething past in dark suit
and Christmas tie?

Unlearn his twenties?
Rise to occasion
like a pond full of goldfish?

I have sent out
a new set of orders
to the cupboard-wowsers,

you are all
blessed selves, on your toes
running before the waves,

identity flashing again
on halfmoon beaches
like a winter sun.

THE COLLECTIVE INVENTION

Wanting a myth for blowing up the gods,
In his flowing silver cape and plastic wings
Captain Melbourne zooms over Lonsdale Street.

His faithful girl assistants screech below
In their custom runabout:
Their T-shirts bear the sweet inscription JUSTICE
And they wear it well.

Admiringly, the crims and their dwarf henchmen
Hustle away,
Skulk underneath overpasses,
Terrified of the sterling Captain
And those just bosoms busy on their trail.

Glancing up from our sandwiches, if we're lucky
We might spot him gliding overhead
In his line of business,
Doing the common good all day long,
A handful of thunderbolts in his shoulder holster.

O Captain Melbourne, man of the wind and rain,
You are our hero,
Save us from northern weakness,
Continue still to zap and biff and pow.

KEEPING ONE'S HEAD

Look over the parapet: parks tuft and burgeon
Smitten by summer brilliance
Where the future lies unmade.

Five miles away a yacht-sail flecks the sea.
I only turned away for a moment
To kiss a warm shoulder, read a book, just stare,

But it was long enough.
When I focused again the universe had changed,
My kinds of music were gone,

Old feelings had passed away,
No one could remember colour field abstraction,
Jane Fonda seemed as old as Persepolis

And people were chemically different.
Forget it. The unrelenting skybowl
Is cloudless and the huge elms drowse.

Look over the parapet.

GENESIS

Then Eve picked the apple
and taking a good bite
learned the first lesson of humanity
which is knowledge, pain and loss,
so that Adam following after
but out of the seeds of love took an apple to eat,
becoming in this our first father
to labour, sweat and face
darkness welling like a spring tide
in narrow straits.

Darkness, darkness in sunblaze,
then a whole garden transfigured
to the serpent's glittering delight.
Look, look at the fruits,
prolific, dangling in mass and tone,
blush on the peach-soft cheek,
cherries like gems, oranges burning,
sex of the pomegranate cracked wide open,
swelling in fig and pear—
that whole estate was suddenly filled
with new discriminations,
with choices to be plucked.

Why was Adam bitterly weak,
reproaching God and woman
like a spoilt child with broken toys
snivelling and whingeing?
Why were his guts weak
when he was called on to make a stand
in this brash new medium,
fast-flowing time?

It only took courage.
Be a man, said Eve;
Be a man, said ambiguous God;
yet Adam snuck away and moped
like a poor put-upon creep
in the shrubbery, in the evening,
scared of the too much light.
Know yourself, Adam.

Why was Adam bitterly weak
with her love, their sweep of knowledge,
work to be done
all over the face of the earth?
Everything was possible
that you could hurl in death's teeth:
the agricultural revolution
slept in ungathered grain.

The serpent was finished, mere reptile;
you break the back of the tiger snake
as soon as look at him,
farmers hang them on fences.
But there was a question for Adam to ask;
why on earth had Eden's God
chosen to make the serpent subtle,
crammed its jaws with language of men,
a trick to cap all creation?
who was kidding whom?

I hate the story and love it,
detesting death, a vast stupidity,
but glorying that Eden
could be smeared with, flashing with, energized
by the first colours of love.
Everything came alive.
The dull stuffy paradise park,
that silly supermarket
frozen away
under Claude Lorrain's stiff glazes
became part of the world;
here a fresh wind tossed the branches,
rotten fruit fell, green fruit waited,
ripeness was all, all transient
as man's quick breath. Things mattered
and love, anxious love
rose and put forth its flags.

THE SHAPE-CHANGER

The first day he was travelling to Asia,
the next day he flew the flight of a wedge-tailed eagle,
the next day he was the gusting wind,
the next a bright campfire
the next he thought of St Kilda on those open, drifting,
    sleazy summer nights,
the next day he was a seal, big-eyed and sleeky-brown,
the next day the little cousin of Death
and the next a writhing snake
or an ancient painted clock with two pewter soldiers to
    strike the hours,
the next day he stood with all the workers, shoulder to
    shoulder,
the next day he grew like a tree, covered with sunlight,
the next he was back swimming off Elwood under the stars,
    many years ago,
the next day he was a yellowish lion
and next the sandy, howling wind.
Being Proteus, he never dreamed at all.

BENNELONG

When I blew to Europe in the big white bird
or nibbled at lettucegreen England
I could never forgive the landscapes

for not being southern.
They were pretty
follies, they were

colourful imitations
of the crazy landlord's personal set
of watercolours.

They had charm,
fatigue and kitsch.
I burned like an old stump.

The big dry bones
are wedged under my flesh
in a personal geology;

leathery leaves are whispering in my guts;
slow brown waters churn through me.
O Jesus, Jesus, turn me inside out.

NEW CARPENTRY

'Check it in here,' he said
squinting against the dayshine.
'Mortise and tenon. Yeah.
Chuck us over the hammer
where it's lying beside you.'
Whacking his chisel into soft splintery wood.

No metaphor
the real thing.

Cutting a piece of oregon to length,
blond sawdust flying out in puffs,
tarring the end
two inches of blackness for mother earth.

Now he drives plugs into
the old brickwork corner—
soft, friable, orange bricks—
and nails the upright into place
with a chiselled gap for lintel
four by two.

We are ready to hang the door.

PUCK IN JANUARY

The moon is full as a goog floating
                        in a mackerel-backed sky
Outdoors, nine-tenths bare,
                        I feel liquidly cool right through
And hear a faroff dogbark
                        faint messages of insects
The warm stars seem to go walking
                        all over my body
Which feels nippy and brisk even though
                        the mind drags seasonably
Sometimes it is quite easy to be happy,
                        fit as a Mallee bull
My very name
                        has wandered away.

PANTUNS

I am the most terrific snob
who can't approve executives
and all the garbage they intone
as canons of successfulness.

In matters of aesthetic flair
I am the most terrific snob.
Don't give me canting populists
and all the garbage they intone.

I dig Mozart and Baudelaire
in matters of aesthetic flair
and when it comes to novels, boy,
don't give me canting populists.

Like every bourgeois radical
I dig Mozart and Baudelaire,
cannot stand TV comedies
and when it comes to novels . . . Boy!

Destruction of the state delights me
like every bourgeois radical
and I, the most among all these,
cannot stand TV comedies.

The small vulgarities affront me,
Gilbert and Sullivan I hate;
destruction of the state delights me,
much the most among all these.

IN THE WILDERNESS

Our eyes woven together
my hand under your spine

your tongue like sweet salmon
darting between my teeth

that skin, can it be yours?
your limbs have entered my own

hair brushes through hair
our colours are only moonlight

my body ebbing and surging
as I reach towards where your heart is

mine is diffused everywhere
the desert is cold—we are warm

I am entirely you-flavoured
your being marches through me

as we drift above cloudbanks
or lie in the world's lap

tomorrow may taste of ruin
but we drink the one soul

tomorrow we have names
we are shaken to pieces tonight

OLD MEN DURING A FALL OF GOVERNMENT

Mortality grows all over them
like a field of flowers.

Their bodies are geologically
pre-Cambrian or such,

gullies and dingles everywhere,
erosion, moraines:

you can glimpse the original outline
the colour of reminiscence

before their eyes retired
behind these fleshy blinds.

The political bloodfield thunders
but its runnel of destruction

does not even reach them.
They are unproductive paddocks

out back of nowhere.
Look at those veins, like poppies,

and that once-ripe hair,
a thin scurf of snowdrops,

and the tears.

NOW THAT APRIL'S HERE

for Peter Steele

Where monotone was, the tips and fledges are uncurling,
        pastelled, peagreen, purplish-pink;
bannerets, dwarf opening brollies of ardent verdure
        change the profiles of shrub or branch.
The hills' faces are mopping and mowing with growth.

There is a hunger for difference in the air,
        new attitudes to sunshine,
goodness of earth rustling to get up and go.

The stillness is crammed with tiny excitable statements
        held on juicy short stems.
Listen: the almond has something white to announce,
        the cherry's pink has green verbs,
all active, and blackbirds are doing their best to keep up.

At home, folded in slow continuous seasons,
        their bluegrey miles, we cannot guess
all this Vivaldiesque agitation of April.

Like the tense buzz of political parties
        that split, proliferate and fray,
the flowers come. Our stock of abstract ideas
        is exploding like a bomb,
confusion firing the mind. The shoes of promise

are crunching round the bends on gravel paths,
        slow, quick, slow and a couple of skips.
Look at the cloud in the shape of a dud guitar.

Berenson, Machiavelli, obstinate Galileo,
        these are your chatterbox hillslopes
feathered and all ajump with continuous growing;
        shapes of republics living or dead
carrying their carnival colours brilliantly through my head.

THE MENTAL TRAVELLER'S LANDFALL

O miraculous blown country
smelling of grass and hucksters' money,

rejoined stem of my sappy world
rising in the bird-riddled sunlight,

streets as wide as oceans full of nothing
but good humour and middle-octane corruption,

enormous anthills plumfull of Greeks and Sicilians
dropped here and there by parenthetical chance,

turfy meadows punctuated by batsmen
and neoplatonic fauna that Eros devised once

and tucked away south of the Wallace Line,
Yoo-hoo, I'm back. Let your furry paws go clap-clap,

roll out the ivory carpets, the spectral champers,
your childhood slides and the tin of cocky's joy.

The dead are alive, they are diamond-coloured dancers
on a triumphal car as the weird bankers

accumulate dung in their vaults (growing votes and such).
The dancers are risen, they are great rustic virtues

and metaphysical decencies, straw in mouth,
they are bright as tinfoil, they burn like martyrs' blood,

they meet the full moon wedged in a yellow-box fork.
They are one's better, baptismal self,

hater of history and motivational sludge,
wouldbe titanic cousin of the sapphire waters.

A sort of amniotic stillness dissolves me
full of mineral salts, trace elements

and the secret circuits of association:
my veins are iridescent

while black dwarves find gold in my bowels,
phantasmal country like a blowing map . . .

INTROSPECTION

Have you ever seen a mind
thinking?
It is like an old cow
trying to get through the pub door
carrying a guitar in its mouth;
old habits keep breaking in
on the job in hand;
it keeps wanting
to do something else:
like having a bit of a graze
for example,
or galumphing round the paddock
or being a café musician
with a beret and a moustache.
But if she just keeps trying
the old cow, avec guitar,
will be through that door
as easy as pie
but she won't know how it was done.
It's harder with a piano.

Have you heard the havoc
of remembering?
It is like asking
the local plumber
in to explore a disused well;
down he goes in on a twisting rope,
his cloddy boots
bumping against

that slimed brickwork,
and when he arrives at bottom
in the smell of darkness,
with a splash of jet black water
he grasps a huge fish,
slices it open
with his clasp-knife
and finds a gold coin inside
which slips
out of his fingers
back into the unformed unseeing,
never to be found again.

GASPARD DE LA NUIT

Are these all many or is it one?
enquiries that revolve,
move the feet and thump the pillow
round about 2 a.m.
with pale blue moonlight splashing on the carpet
and a mopoke moaning somewhere
among the pokerwork hilltop trees,
odd cars bent homeward
distantly.
One or a medley,
night, nuclei.

Our galaxy has no point at all
nor do the others
gleaming down
on our marriage, bills, piano,
on Parliament House,
T-shirts and ammoniac nappies
lodged in their chink in space–time
for some directionless reason
soon to be rubbed out like a blackboard lesson.

Sleep now,
dreams are much easier
and cheap as chicken broth.

MIND

Along the face
of this baffling space–time construct
an illusory straight line goes ticking
fringed with words

in its long push
toward continuity.
Somewhere below
a chafed sea sobs

washing up the monsters
through tanklike dark
and a thousand fragmentary
shades of green.

The tank's edge is flicked
by things of the world:
frond, blade, loop and wing,
tiny ecstasies of light

all laying claim
to a genuine status.
Odd packets of language rise
to acknowledge them

while displaced waters ripplingly break
on pain, loss or breakfast.
The plot is set on a small star's
fortunate planet

which we (so to speak)
entitle Here
trying to make
a map or globe of sense

with a child's
original box of paints.

ABHORRING A VACUUM

Pearlmother dawn. It is fairly true,
'The mind divests itself
of any belief in the mental'

but my slept fragments
fall back together into a shape
doing things with cutlery.

Outside the pane, frondage and dewdrop-cluster;
tiny birds in their twelve-tone clamour
recommend continuity;

no one I can see observes me,
who fade like blown steam.
Dozy-dim, I battle back as

the fiction of personality
nimbs me for a moment like
slant light through a tram door.

It passes. I am lived
by who knows what, the gene's blind way
of making another gene.

Whatever has been writing this down gets out
from behind the wheel and
walks away.

NUB

Gro-ink. Kopita, kopita, ko-pi-ta . . . *konk.*
Were it not for the fact that I'd recognized the brickwork
of the previous station, coming out of my daze

I'd not have known. Out. I ran a whole block
at dream's peculiar slowness to catch my tram;
grey blocks of a city abstractly slid.

I came to the upside-down house,
its walls were folksified with Virginia creeper
and it looked as cheerful as a month of Sundays. I went in

being bent on finding the future, the logic of which
led me hotfoot up the stairwell to mother's bedroom
where, plumb on her chest of drawers, lay a single jewel-case

of tortoiseshell. Snap: open: there lay the future,
a perfectly polished chromium steel ball-bearing
round about five-eighths of an inch across.

I stared, closer and closer into its surface
so that the room, my face, were warped and bulgy
as through a fisheye lens, but rather fun.

In such a concentrating situation
what can you do but stare?—that's what I did
feeling immense waves of sadness in my legs

and chest all the while. I contracted the world's pain.
It was then, smaller at first than a bee's foot,
that something drew in closer over my shoulder,

an artefact as much as a consciousness,
jointed, threaded and cogged, full of spindles and heads,
a bronze-and-silver toy of elaborate construction

with a sharp weather-eye open. It knew the game.
It was the future. It gave away nothing at all
but a friendly tap on the shoulder. I shut the box.

EXIT THE PLAYERS

It is over. Self-mortifying Hamlet
gets up and puts his sword back on the shelf;
Laertes, unbloody, shuts his manila folder;

the Queen, who moved two separate amendments,
thinks Yorick played the fool offensively;
Horatio had the numbers this afternoon

and buffaloed through a policy decision
the Ghost would not have stomached;
Denmark is feeling just a wee bit whacked.

Ophelia's hair is dry; she didn't say much
but wouldn't mind slipping over to the pub
with Rosencrantz—why does he have to keep nattering

to the King? It's drinking time. Stiff Guildenstern
will not forgive Polonius and, to boot,
has lost his biro somewhere round the place.

It could have been worse. The dead all kept their tempers,
Gertrude cracked several, Hamlet one good joke
and they got through the whole agenda, perhaps because

Fortinbras is still on sabbatical leave.

THE BITS AND PIECES

Swift's discovery, fundamental for art, is that there are no uninteresting objects in the world so long as there exists an artist to stare at everything with the incomprehension of a nincompoop.
—Andrei Sinyavsky

Artichoke

Children's drawings of trees
approximate you
with your striated trunk
and dense neat head
inflorescent
of purplish scales or flakes
for our buttery meals,
seeming a sort of edible pinecone,
a green knight's club
or else an absolute rose.

Banana

The lordly nutritious banana
is peeping over the crunkled rim
of some old pottery bowl or other
up on a sideboard, at vantage;
look how his nose is black,
jetblack as lovebites that blotch
the yellow hide of his throat,
tropical patriarch
knowing perhaps
the new moon four days off.

City

When you enter some part of a city
that is completely strange
what excitement leaps inside you,
what delight,
milkbars and palms whirling like paradise,
each trim corner the colours of promise,

verandahs dancing,
stucco ready to please the eye.
I am good fortune
I am your future
I am the names of forgotten men,
it seems to say.

Doiley

But the lacy components
of this cappuccino-coloured
map of a uniformly
expanding universe
are crooked stars,
explosions with their eyes poked out
or everlastings laid
in ranks of flat little boxes
promising a mathematics
none of us can read
or a seductive net
containing nothing.

Emus

It is
particularly
the particular way
they come
stepping
warily
down the path
in dark
wrinkled
stockings
and shabby
mini fur coats,
their weaving
Donald Duck
heads
ready
to dip
and snatch

your ice cream
that appeals;
that
and the way
they browse dumbly brown
in cattle-paddocks.

Foot

Scalloped with toes,
flat as a hat
and pink as any nudity,
these ridges
and these shelly plates
will superscribe its crudity.

Fresh out of wool
as a hermit crab
it creeps across the carpet pile;
a mirror-sister
marches past
then falls behind it by a mile.

The range
of possible expressions
is limited to flex and kick.
An ankle canters
on this mare
whose littlest buried bones go click.

Galvo

Those large gaunt shearing-sheds
on the long tilt of
                              halfshaven hills
are built of it:
they are pastoral cathedrals
of dull, rippled grey
or steely new ships of the soil
breasting against another winter wind

that fills the pines with whirling blackness.
The rustbrown battered stamping-houses,
dark ovens of history
                in the summer mountains
were made of it
when trolleys ran with quartz and gold.
Loosened sheets bang away
                in yellow Januaries;
dark rotted pieces
                lie in the creek's edge.
In the backyard
a creepergrown green dunny;
at the township's edge
threatening humpies of the defeated ones.
Galvanised iron
                our modern thatch
for a long time you wore blue slogans
for Doctor Morse's Indian Root Pills
near this or that sequestered cowyard.

Hair

Heaping, glistening, a waterfall
light enough for the breeze to leaven,
this all too human foliage
to fill a lover's hand, a coffin
or the red eye of St Paul.

Ink

With scratches,
splutterings
and with slow
determined blots
your thick fingers
learn to become learned
in it:
how to master
(Master Pupil)
the tyrannous

steel nib
into a painful
parade of copperplate
with no crossings-out,
into a decent hand,
the linear
life of the mind.

Jasmine

A presence—how can you name a smell?—
heavy, hazed, like peaches and honey perhaps,
spreading over the lawns and lanes;

look to the flowers, dense on their dark-green
pointed leaves and upcurving tendrils.
They are fivefold stars of white,

they have crooked points, they are
spokes of a series of wheels turning clockwise,
there is a touch of yellow inside the corolla.

Sweetness of jasmine,
it bends up from the rickety paling fence
by untended beans and caulies.

Can such floods
of scent
come from those frail starfish
bunched high on the tennis-court wire?

The gravel swoons,
the lobs drop back.

Kelp

Slowly it blackens
on the yellow shore;
a hardness thickens
more and more

in leaf, bulb, flange
and rubbery stem
along the fringe
or scalloped hem

of surf-surge. Time
turns all opaque,
including these
straps, grapes, trees,
fan, tress and rake:
gone their soft prime.

Lava

With glaring edges, lava crawled over the low
smoky plains, and stayed; slowly antiquity
chipped it and crumbled it.
                    Hardness decayed into loam
plumfull of bluegrey boulders or soft in the flank
of dawdling watercourses. Drifts of dark folk
came and went. Winds out of the west
hurry their squalls gusting over the sheep
new, canny folk know how to fatten and shear,
or change for cattle. Their hollow churches
bear the bleakness of that old hard basalt.
Violet, smudged evening: a vague dream of volcanoes.

Marmalade

in its tropical rock-pool fashion
holds quaintly shaped odds or ends
teasingly in sweet suspension
all suffused with light,
a saturation of old gold:
amber, chartreuse, blond or rusty.
With a waking touch of piquancy
it weds your grainy toast
to consummate an aubade.
Powerful, insinuating, yummy,
it is civilized Scotland's grand gift
to the barren porcelain of Europe.

Nectarines

From such rough barky trees
and crinkled crescent foliage they come
bearing the press of years
or boyhood's tang,
sharp, heady, themselves,
not this nor that
keeping their seasons happily enough
and turning one red cheek
toward the embrace of sun
or sharp white teeth.

Opener

is an astonishingly slim
metallic biped,
that is if you call
Captain Ahab a biped;
knees together while
his two teeth bite down hard
if you twiddle his ears round and round
and he skates round the edge
of a silvery rink
on his head,
in his frustration
biting right through that tin crust.

Passionfruit

used to be plump and glossy
but his mackintosh has shrunk
as he sucks in his cheeks
while somewhere inside the room
he is giving you the pip
with a wonderfully sour sweetness
like last year caught in a daydream,
the tang of paradise.

Quail

Quail in a beer-garden
brown behind netting
shy as the long drinks pass.

Roses

Oystergrey
thick rain off the vertical
on convolute roseforms large and small
a hundred kinds of blush
carrying fossil emotions
a thousand years or so
dangled from each wet petal.

Tenterhooks of rain gingerly depending
from that which is.
Theatrical trophies
(melancholy queen)
bobbing on their spiny butchered towers
this densely implicated origami
for whose soft host of kindred colours
language has no words
                    (but surely there are words).
Curvacious, polygonal, rich heartfilled cupcakes
when I bend down to the table
they slide away
into their rosy selves
                    the one and the many.

Oystergrey
this hydra-headed showerfall
faintly clouds the score
of unbelievably resistant shapes and types
all of which mean rose
and are.

Skies

At Boolarra someone stitched the clouds
                    to the edge of pillowing hills,
the swamp harrier, maybe, or the blue crane,
where dams like sightless eyes look up at stars or the sun
rolling above us blindly over and over again.

Telephone

Why does it drive me up the pole?
Perhaps my hankering for the whole
man, woman, child, in conversation
and not this echo of depletion
which murmurs through a diaphragm
following some branch-line tram
of thought in disembodiment
piles fuel up all round discontent.
Give me a presence, here to hand,
that I may view and understand
but not incapsulated voice.
Bad magic, this; a slice of noise
far from the circle of oneself.
(I wish it neither wit nor wealth . . .)
One's truest image stands close round
breathing upon Tom Tiddler's ground
and cannot be distilled for sound.

Underwear

On the whole they have been much neglected,
especially by writers, these necessary smallclothes;
figuring as Linen in the eighteenth century
and vanished completely of course in the next,
in our own time they seldom score a mention
except as trashy erotic accoutrements
to be 'impatiently ripped away'
in the more boring kinds of fiction.
Give them their due: they do yeoman service
to our crotchly comfort, our body temperature,
cradling of boobs and ease of movement.

Whoever has fidgeted through a bad day
in awkward underpants knows the difference
they make to our psyche, and learns to praise
a technology which has learned to create them
light, resilient, airy and snug.

Veils

I heard the suburbs talking to the stars
in a dialect I could not understand
full of dew, amber lamps, familial mortality
and Brennan's Wanderer fuming along the paths
of himself as the first pale veils came down,
dovegrey, dovewhite, veins full of cottonwool
threading our crescent and the sometime creekbeds
like snow in the mind
               tulle through a nocturne
                         blurring on glass
till the sharp starchart vanished from my ken
completely.

Washing Machine

    Behind a glass pane
    the wardrobe's wet personae
    play at vertigo.

X

    You know that he is passing by
    just on the barely other side
    of marginal experience
    by heartstop, a breath on your neck,
    dim whiffs of garlic and old iron,
    that ceiling of your mouth gone dry,
    all fruits and leafage petrified.

Yams

As the crammed globe
rolls eastward on its axis
into the dark of history

carrying its usual load
of tyranny, art, gods, muddle
and excruciating pain,

it pleases me to think
of the humble yams
stacked in their market stall,

ugly, knobbled clubs
of mediocre taste
lacking even

that noble mythology
which has grown around the spuds
of Sir Walter Raleigh

and a million bags of chips.
The yam has no glamour at all;
it merely lives in the world.

Zephyr

Yellowish, crumpled, frail,
the leafboats lie
on water black as hair
under a marbling sky;
the secret wind moves through them
unseen but visible.

PANOPTICS

It was another race
with a slender hold indeed
on the lintel of prehistory:
a graminivorous people
gentle, bandy, greeneyed,
who left so little mark
on the tofts and crofts of time,
their adzes unremarkable,
shards pitifully few
and their passions blown away
like the long lavender wind.

They gave myth to the stars,
looked into that unbelievably
complex mosaic of
diamantine fires and
called the clusters by names:
there were The Kiss, The Groin,
The Phallus and The First Dream,
Climax and Dayspring.
The Hip, The Labia,
The Great Exploring Tongue
and Semen of Gods.

It was another race,
gentle as rosewater,
libidinous like kittens,
who coined such poetry,
who got dusted away by process
to leave not even
their names on the starry chart
or in the sharp sand.
A new taxonomy
keeps all the stars in place
leaving our dreams alone
and the gone magicians.

PRACTITIONERS OF SILENCE

So much that happens happens in
the gaps between, the spaces,
so much meaning cries aloud
from what you didn't say

or else he didn't
or she was looking away
thinking the quarry on the hill's blue flank
had a kind of truth, too.

Take those exchanges: 'G'day.'
                                    'G'day'.
'I been thinking.'
                    'Yeah?'
                                    'How're the sheep?'
'Not bad. Been dry, though.'
                                    'Need a few more?'
'Maybe. How's the truck running?'

and so on, at the gravel margin,
vibrations hanging in the air
like a black angel with a fiery sword
or a mortgage spouting blood.

Between the stanzas, under the words,
whiteness like chalky bones
as orange clouds float over, one by one,
rhetorically immortal.

SACRED RIDGES ABOVE DIAMOND CREEK

for Les Murray

I want to make some kind of gesture of alien response,
response no longer alien, response finding its feet,
salute with my feet and my hands and heart to the totem beasts,
by whose names this district was patched and pieced like a tartan,
to echidna, formed like a child's drawing of an explosion,
to dingo, long driven away by Mr Fox in his red hunting coat,
to platypus, the shy, the watery secretive,
to magpie, sweetest yelp of the morning,
to sober-suited shrike-thrush, epitome of musical variation,
dour tortoise like a stone
and vibrant kingfisher smart as paint.
These are thinskinned sandstone hill-rows
remaining shaggy with yellow box and stringbark
despite their undercoat of heedless houses,
and the gutters still gush to tanglethroated creeks
so I pay my spilt tribute to all those neglected totems,
to clever-tailed possum with his unspeakable voice,
to willy wagtail in flight, a whole symbolic system,
to kangaroo, whose name at least we dared to retain here,
to square koala, cosily hunched no longer in manna gum forks along
    Main Road,
to black bream and snake
to kookaburra, raucous herald of fire's daily return,
to the wild cat, whom I do not know,
to the 'furred and curious wombat',
to you all, primal kin of the region I choose to live in,
each genius loci now displaced by incomprehensible names and grids
though the cunning birds, secularized, remain,
I yield you primacy now,
you and those adamic makers who prowled around these scrabble-
    shingled hills before me,
whom we may choose to ignore at our own cost,
hail and farewell, but then, after all, hail.
Yours is the first magic.
Yours are the names of things.
I cannot divest myself of this curious tongue
but I lay down my arms.

A STONE AGE DECADENT

Uh.
　　Uh.
　　　　These errant stripes of sun
That feather in play across my legs design
Transient ochres, ripples which the Sea
Has lent the air. I seem to like it here:
My tribal brothers work back up the stream
For tucker; one small knot of womenfolk
Go gathering shellfish where gold sand meets rock
There in my middle distance. They compose quite well,
Sun-burnished nymphs and mothers with dry tongues.
Under these casuarinas on my slope
Of sandstone and soft needles I may hold
A laid-back peace, keeping my cavernous head
Well stocked with pictures.
　　　　　　　Mm . . hmm. Let them retain
Their rules and moieties. I disturb no one,
Neither affront those boring Bluetongue rules
Nor trespass on Echidna's blunt decree
Here where a sea-breeze lightly lifts my hair
Flavoured with faint salt. Tribes are wise enough,
Let them think me no-hoper if they want to,
It troubles me no more than bushflies do
While thoughts waft up this hill: from here old Sea
Is crinkle-turquoise—rumpling, ruffled, white;
Gulls and swallows thread it.
　　　　　　Uh.
　　　　　　　　Sheer solitude,
Watching these topmost branches bar sky's wink
With their shockhead tresses. This is my secret
Adaptation of totemic ground,
Sequestered high, brown half-shade where I lounge
Sending my spirit out to meet the Sea.
Fly, fine colourless bird, on thinky wings,
The words we use are only the words we're given,
They do not like to hear me saying that,
Preferring old songs, with their boom-bam-boom:

Here the big wave runs upon the shore.
Here the spray blows up and up like smoke.
Daily the shellfish, daily.
Young girls gather foodstuffs by the white sands.
At night the seagull has stopped crying
Daily the shellfish, daily.

Totem and law, laws and restrictive totems,
Banal, sublime, bestial, that's how
My fellow tribespeople make out the world.
Practical sure enough—the food comes in—
But bone between the ears; at their sheer best,
Witness down there, say, brownbeautiful yet dumb.
Motes dance in light-slant just above my skin
And the glow-filled Sea flakes off her thousand colours
In tides of mystery.
        Listen.
                Gull and currawong
Sound their antiphon. How fast the shadows
Lengthen on sand, coarsen the hill-textures.
There is pleasure in it all if you sit still.

I do not think they like me very much,
Not even Moama with her small round breasts,
Scrub of light curls, pool eyes, fastmoving limbs
And buttocks I could cup in these two hands
And then . . . Oh-oh! It's very much too pleasant,
That's to say painful, this line of thought:
Body responds. There! She stoops at the rocks.
I see only a single cloud today,
Thin, flattish, grey-white, drifting above the horizon,
All else says blue meets blue. And I relax
On springy casuarina needles here, my den
With a view. A spinebill's vivid uniform
Flashes to flowers a little down my slope.
I flex, reflect, withdraw. Ah, me. We all
Must learn in a line of days to wither up
And die—or else die first. Just like the scallop,
Mussel, periwinkle, any living thing.
Ahi! Know something now. Am I a fly?

Here the spray blows up and up like smoke.
Daily the shellfish, daily.

Peer closely at these jointed leaves or branchlets,
Green fingers of slender skeletal hands
Knobbed with small tan knuckles. Just to stare
With care at this or that makes world seem good,
Be it spiky conelets or multiply-scored bark:
I like it here. Those women on the sands
Make up a dance that fits a larger dance,
The bay, the hills contribute to my joy
As I do nothing. Ha. Yes. That's my game,
My hunt for needful store of images.
Lovely, yes, but what substance underlies all?
What might all change mean? Are we like shellfish
To be shucked and eaten? Why does the great sun set?
I wonder how we tagged these words to all.
Life is more than animal grease and ochre.
I well might fall asleep . . .

THE AMOROUS CANNIBAL

Suppose I were to eat you
I should probably begin
with the fingers, the cheeks and the breasts
yet all of you would tempt me,
so powerfully spicy
as to discompose my choice.

While I gobbled you up
delicacy by tidbit
I should lay the little bones
ever so gently round my plate
and caress the bigger bones
like ivory talismans.

When I had quite devoured the edible you
(your tongue informing my voice-box)
I would wake in the groin of night
to feel, ever so slowly,
your plangent, ravishing ghost
munching my fingers and toes.

Here,
        with an awkward, delicate gesture
someone slides out his heart
and offers it on a spoon,
garnished with adjectives.

THE WELL-DREAMED MAN

And in my dream I woke from the one I was having,
a dream which was nothing more than mere,
began to walk up the S–bend gravel track
which I first pictured some eight years ago,
lay down in thick sweet grass like a boy in a painting
with my straw hat over my face
and began to enter the dream which includes all dreams,
which proves at once particoloured and spacious
with a grainy feel to it like old floorboards
but gets confused with a black-and-white film I saw once
or think I saw but maybe only dreamed of,
so vivid prove its memory traces tonight
while belated cars groan fast along Brunswick Street
less than a block away
with hearts in their mouths
as though they had something big to offer knowledge.

THERE

At the bottom of consciousness there is a clear lake
The waters of which throb ever so lightly
(Like the bodies of lovers after their spasm ends)
Throwing dimpled distortion across the rocky bed,
Greenish round rocks, the size of a grapefruit, say,
And through these cold waters fish are swimming
Seeming quite continuous with their medium
As sexual love flows directly through God.
Here water moves the slubbing barabble of language,
Gust and pith, cacophony, glossolalia,
Gift of the gab and purple rhetoric,
Moaning in rut, scream, snicker, and the rip
That is sheer pain.
                            Yes, these are of language
But not yet *it*. They are the pool,
Its diamonds and yabbies, ripple and scale,
Insatiable glittering . . .
          I'm afraid I don't know what paths
Lead up from the pool to where I think and talk;
By what stony track with landslip and synapse
Distracted everywhere, choked with scrubby thorns
We got to where we are. Conscious.
                            Aren't we?
Oh hell, we seem to think we understand:
When I ask at the ticket-box they sell me a ticket
But I do not know what the recently dead will ask us
When they walk through the scrub again like sunbeams.

THE STARLIGHT EXPRESS

If you can slip

under reality's near edge

to hear those other things:

crickets tuning up

swash and drag

clock all bustle

but a bird somewhere

and thin bright stars

whisper secretly

You know that.

I know it, I riposte

but please do not let them

A moment of pure silence

like the scepticism

your fingernail

you then begin

a child's yawn offstage

in bootblack hedges

of immemorial surf

fridge idling away like a tired hound

enjoys an idea

in their savage beauty

'You must die.

You have to die.'

uncomprehending

burn my body.

grips all nature

of angels.

A GLIMPSE OF SHERE KHAN

So the jungle
was pard and barred,
reflexive with shadows
patterning our path
as the coir saddle rocked slowly.
We were all eyes.

Then Bashir, mahout,
straddling behind the beast's ears
gripped my wrist hard.
'Tiger,' his teeth hissed
as our elephant began
a long, strong bout
of shuddering through and through,
head on to where
stripery had been and gone.

Around, back and up right,
canting to cut him off
in some douce clearing
and—
     HIST!—
          so we did:
all aglow now, his yellow
astripe, approaching, grand,
he had us heart in mouth.
We stiffened. He stood,
shade on his hide,
the whole dry glade
gone tense. It seemed an age
then, flick, he was lost,
filtered through thin jungle
slotted in space-time.

GENIUS LOCI

When I can't sleep and prove
a pain in the neck to myself
I will sneak downstairs, dress up warmly
and squeeze into whelming darkness
or piccaninny daylight, where
I may just glimpse at a corner
one of the Jika Jika slipping away
lapped in a possumskin rug.

I will hurry like steam to the corner,
ever so much wanting to say,
'Hey, wait. I have so much that I . . .'
But there will only be
broad street, creamy houses, dew
and a silence of black shrubs.

Maybe if I got up
a little more smartly next time,
got out on the road quick,
I could sneak up closer
on that dark tribesman in his furry cloak
and ask him . . .
                    oh, something really deep:

something off the planet.

SONNETS TO THE LEFT

I

What do I get from progress? I rejoice
In antibiotics, the dental drill, clean drains:
Perhaps a little bit in aeroplanes.
Deutschegrammophon and His Master's Voice
Pump a magical suasion through the air
So that we all have orchestras to hand
Or history on deck like contraband,
Bartok and Bach concurrent everywhere.

Slavery's dead. An evening chill now falls
Rather more golden; autumn shadows flit.
I wander round these white functional walls
Trying to find a frame in which to fit
Large things that progress bundles out of sight:
Grief, awe, terror, transcendent light.

II

(i.m. *Judah Waten*)

As we grind into the worst decade for two
Or three or four, I swing around and see
Your bulky, suited, Russian form push through
This or that minor bookish jamboree
Smiling, and think of old hostility,
Those years when I watched you hard for Stalinism's
Cloven hoof and you (I'm sure) marked me
As bourgeois formalist. Time burns the isms.

Now Toorak and Balmain contrive to read
Marx as ur-text, floppy aesthetes delight
In cushioning off him too. The game is bent
Till we've become old colleagues with a need
For shoring words against the tide of night,
Praying the slow bitch History might relent.

75

III

There used to be a time, as I recall,
Where 'Left' involved the proletariat;
I like to think that might be where it's at
But fear it's grown up theoretical—
Or else gone troppo and turned Libyan.
For years it strove with muscular intent
To keep the ALP from government,
But we've now crowned that gross amphibian.

Soft acker vikings cross a phantom sea
Hoping to put old systems to the axe
And loading up their vocab, trend by trend;
Joying in papery perversity
The deconstructors lie down on their backs
And let our Marxists rape them, in the end.

IV

The writer depersonalizes his dreamwork,
A single being in search of voices,
But change is learned from the outside world
Preoccupied with drink and kissing.
Unhappiness is the funniest thing,
It is a play about the ego
Needing to choose a lifelong project
Through tea and comfortable advice.

How can anyone be called guilty
From the perspective of the trapped
And struggling fly? Perhaps the itsy-
Bitsy is proof against despair,
Knowing between the fork and the knife
All our beyond is in this life.

V

We look edgewise
Through this our dishy galaxy
And see thick jewellery
Spattered across an empty sky
But our damp cries
Balloon at last through history
Which does not leave for personal foolery
So much as a footnote or sigh.

What'll we do with the mystical
Now that red buttons think beneath our thumbs
And secular hope is built upon a rock?
Turn to the fantastical:
In furious rubber last of all there comes
The hooded cyclist with the duck-beak cock.

VI

Look, mate, you roll new brands of foreign cringe
With which to snow the members of your branch.
You are a proper old conceptual stinge
And where you gesture at an avalanche
We only see the dreck of magazines,
Half-baked ideas, half-read, near understood;
You reckon you have proved you're full of beans
By squeaking, 'Foucault', through the sacred wood.

Cling to your books and views of bin-end booze;
That way you won't get dogshit on your shoes
Tailing along to popular festivals.
Look,
        something real is happening out there
And when you finally notice, then, I fear
The feminists will have you by the balls.

VII

Tell this to Yale and Paris, tell Pan Am,
Bloody imperialism never ends;
However much Big Brother dubs us friends
He'll roll us up and dump us off the tram
At the drop of a hat. Ah, *dump* . . . yes, dump:
That is the linchpin of imperial trade
When the time comes to slip the masquerade
And knife your friends under the doom flag, SLUMP.

Australia first, and last, and in between
Australia once again, is playing safe,
Even for us, the rigorous aesthetes.
High art begins from kids' play in the streets
And local heroes earn the long last laugh:
Ocker, metaphysical, obscene.

VIII

Through a green, social afternoon
As to and fro the emptied bottles roll
This question rises, bleary, out of tune:
Can a fuckwit have an immortal soul?
When shall the vulgar court the lyrical
And bullshit fertilize the laurel crown?
When will cute images delight us all
With ratbag stylists going on the town?

Only when stories happen on a seam
Whose gist gets memorized from north to south,
Swimming through mythology like a dream:
The dingo with a baby in its mouth.
Fuckwit and smartarse, trendocrats and folk,
Nothing unites them like a dingo joke.

IX

De Tocqueville said (he really didn't say)
That in the art of democratic nations
The brittle courtly forms will shred away
And be turfed out by more offensive fashions.
O.K.
         Sing of the footy, sun, red wine,
Green surf that batters on this ridgy coast
And then, as redolent as turpentine,
The spirit of the place: our compound ghost.

I like this rhyme, so how about a toast
To that bold Rhadamanthus, tough Jack Lang,
Our patriot whatever (oops!) the cost,
Whose name, returning like a boomerang
Led to this rough platoon of sonnetry
Affirming Land Rights and democracy.

X

Sweet as a clarinet the dawn returns;
I hone my slang aubade, damn sure that our
Priorities are not the Comintern's.
Land Rights lie deeper in the soil than power,
Being the native debt, profound as trees
Turning metallic leafblades to the sun
In morning's natural order. Pieties
Deem that this act of ritual be done.

News twangs a bleeding knell for every age
Or else a snotty-grey decline, unless
We've laid out clear proposals down the page
To crank ourselves out of a fiscal mess,
But if ideals have something left in hand
Tribal sites will peacock a beige land.

THE MIRROR STAGE

I met this little girl
One lunchtime, years ago;
Well, she was a widow then
But she had been a little girl.
We used to call her Missy.

And she suddenly spilled the beans—
The widow, that is, in our Staff Club—
That back when I was a boy
They were told not to tell me
My father was killed in Burma.

He was missing—four months or so—
Painfully trudging through jungle,
Shot at, magicked across deep rivers.
When the Japs bombed his last jeep
They blew up tins of pineapple.

My father had not been dead,
Just on a long, slow detour.
He savoured three years more in Asia,
The vine of his life, it flowered like wonder,
Never to be tasted again.

Years of mild swashbuckling, miles
Of Indian wanderlust:
A war like technicolour.
Now he is burned, and mother
Is burned, and we've buried my aunt.

My limbs are being cut off
To make me hobble faster.
I am waltzing on their graves
Like a sunstruck hatter,
Like Indian ink.

STARDUST

To a smell of water-vapour and wood fires
Walking by night, my breath allegro
To feel it's all not worth a cracker . . .
But how could the universe have meaning?
Would the stars be patterned differently?
The seasons vanish, or come on faster?
Would there be an End?
Perhaps we wouldn't require any sleep;
Maybe we'd no longer have to shit;
Or one radiant mathematics
Would show up trimly in everything.

Everything is just as bright
As the hollows are Indian ink
While my bones go wandering chockfull
Of a crushed silver.
These paddocks have all been marked out
With expressive diagrams
As beyond the highway ribbon
Big waves crumple and bang.

But when you ferret after the meaning
Which a universe could be hoped to have
(Oh dear, yes, one has a cold,
Or has an exam the following day,
Or on occasion has an erection,
Or they have a holiday shack up the bush)
You stick at a sort of spatial problem:
Meaning is only a bundle of signs
That parallel and light the real,
But would they then be *in* the real?

Pan has left the odd footprint
On somebody's wet lawn
And his hot metallic stars
Are doing the rounds of my arteries:
I can even feel the moon
Down in my quicksilver groin,
The drypoint shadows
Falling across my brain.

Then signs are doublewise at once,
Being inside and outside what they picture:
If not, they're simply beyond our ken
Like God's hand moving among the stars.
We must find a little enclosure for meaning:
It needs living room, like a dog, or a student,
And won't be satisfied with my brainpan.
I hope we can find a cosy gap
To bed it down in after all.

In those juvescent nights of starshine
When I knew what the future signified
A full moon could
Shake me with stony horror
But now it hauls me into
Pure aesthetic compliance
As a pinecone rears over its shadow
On the concrete pathway.
Such fallen silver as this
Leaches down through cracks in the earth
To take the place of marrow in ancestral bones.

GOD

That is the world down there.
It appears that I made it
but that was way back,
donkey's years ago, children,
when I spoke like a solar lion
beguiling physics out of chaos.

I spun my brilliant ball in air.
Such thought was new to me
though I had not guessed at my lack
in the old indigo days,
children, before you fell—
to use a technical verb.

It is full of beautiful flair,
a jewel and a garden at once,
bluish-green with the track
of silver engraving its veins . . .
Shit, but it's lovely
and no end of trouble at all.

Children, it once was bare
of all salacious language,
of goats and bladderwrack,
of banksia trees and wrens.
I endeavoured to bring it up rich.
I reckon it's my museum.

I gave a big party
and the name of the party
kept slipping clean away
from my wooden tongue
but I reckon it was
called history.

Some honoured guests
took off their names
or left them impaled like scarecrow rags
on my staggy front hedge.
I thought of it as being
a party for my son.

THE LORELEI

Some have seen her by the highway
    From Bordertown to Nhill;
As they leave a purple sunset behind,
    There she waits, quite still.

She's been spotted down in Gippsland
    Or along the Great Divide
In the shadow of a redgum
    Hitching for a ride.

Straw-blonde, in a windcheater
    With frayed and faded jeans.
Some think she's in her thirties,
    Some in her late teens.

You are driving towards your future
    Doing eighty down some road
When there she waits on the grassy verge,
    As motionless as a toad;

You ask her if she wants a lift;
    She smiles and sits in the back.
She's only got a handbag
    And a smallish haversack.

You exchange a few small pleasantries
    Then the conversation stops
And all around the narrow highway
    Comforting darkness drops.

Another dozen miles and you feel
    A pricking of your hair.
You look in your rear-vision mirror.
    There's nobody there.

There's nothing at all on your back seat.
    By now you're spooked as hell:
We suggest you cruise on carefully
    And stop at the next motel.

THE THING ITSELF

The important thing is to build new sentences,
to give them a smart shape,
to get acquainted with grammar like a new friend.

One rubs down syntax
into a coarse familiarity,
such foreplay as closes down all thought.

Were it not
that the undertaking is too mannered
(as gnostic as a shower of rabbits),

I would like to go right back,
devising a sentence
unlike any such creature in creation;

like nothing on the planet:
a structure full of brackets and cornices,
twigs, pediments, dadoes and haloes and bells,

full of nuts, butter and flowers!
sinewy, nerved,
capable of blotches or of waving hair.

That would be a sentence to really show the buggers,
like a cute
new thing

or like a tree
recently invented
by some utterly brilliant committee;

it would glitter, articulate,
strum and diversify.
It would be the thing itself.

THEY

Where have they gone? Somewhere ahead of us
in a meadow like the square root of minus one—
infinite pastoral; pure interstice—
where two objects can browse in the same space
and history leaves not even a snowflake's print.
They have passed through darkness into a radiance
which we cannot know and they can't comprehend
but which does not remember the griefs of our world.
The pain is cauterized,
                                    the atoms dispersed.
Body is no more body, nor is it soul.
They are now at one with a nearer face of the All.
Lamenting them, we weep for ourselves.

AN ELEGY

Everything turns out more terrible
than they had said, or what I thought
at midnight they had said,

but the dark marks
tracking across clean snow
way down there must be people,

that is,
if anything on earth can be human
when eighteen storeys below,

so that I wish again
it were possible to pluck my son
out of dawn's moist air

by the pylon-legs
in that dewy-green slurred valley
before he ever hit the ground,

to sweep under his plunge
like a pink-tinged angel
and gather him gasping back into his life.

THE LIFE OF IDEAS

i.m. *A. F. Davies*

Aloe, agave, portulaca, prickly pear,
How these remotely anthropomorphic shapes
Gathered around us in their martian rig
As we walked numbly through an afternoon
Of lessening grief.
                    The eye delighted
In such a weird fair of inflected shapes.

These were the forms I once made up and drew
In backs of exercise books during Latin:
Perhaps their freakishness is vaguely classical.
Rounded up, fed here in a garden of science,
All these pumped-up elbows, clubs and phalloi
Articulate a system clear as Latin

But which I cannot do much more than glimpse,
Reading these plots as barbarous inspiration.
Yes, I'd pillage these monsters for a drawing,
Preferring not to know, in order to
Sneak up on knowing from another tack,
The calligraphic dance of inspiration.

The Life Force has come to a mardi gras,
Stagey bumpkins dressed up in quilted gear
Stuck full of arrows like Sebastian,
Or else ratbag knights from an amateur production
Of *Murder in the Cathedral*, pissed as newts.
How could nature have tricked out all that gear?

Bearing in mind (whatever may be mind) that
Language is the language of languages,
We ought to learn from this taxonomy
Something at least. Such plump allusions to
Plebeian cabarets or rustic orgies
Hide symptoms of the damaged languages.

Damage is where we start from. We survive
Displacing down a value-gradient
The itsy-bitsy fragments of our childhood.
The signified is all that is the case
But what rich slips got buried down the garden?
Grease, grace, gravy, Grandma, gradient . . .

Just naming them will conjure something up,
At best a truth: whatever that may be;
Say, something planted in the numinous
On which the sun leans with particular grace
Like it would on a dancer. What green piercing music
Coerces our shards of happening to *be*?

It is raw grief can lock us into process
The linkage also grown from the forgetting
As uncollected plants may haunt this rockery.
Some of these succulents are jokes about
Utter non-being—
                    I don't know why I said that.
I feel dreadful. I simply need forgetting.

We build from what we're given early on,
Keeping a child's original box of paints.
Freedom finds room amid taxonomy
And calls renunciation civilized,
Though barmy Nietzsche felt that pride and wit
Could reinvent the colours of the paints.

Language is limited but inexhaustible;
Our bodies are the grounds for metaphor
But forebears tell us how to name our bodies.
The double-bind suckles a double mind
With which we wander through Botanic Gardens
Negotiating them by metaphor.

Names for these shapes?
                    Why, anthropoid,
Globulous, falgate, serrulated, drab,
Robust, rupestral, pyramidal, plump

And acutifoil. When we get a grip on the names
We take things into our mental block of flats,
Setting them up as crudely green and drab.

In this prodigious picnic of the cacti
Departures from a norm are the grand prize.
A new idea is always faintly monstrous,
Its novelty being what constitutes
The warp or bulge. It is the spanking new
Reversal of an axiom that we prize;

It is the opposite that's good for us,
Taking the dog of habit for a stroll
On the Big Dipper . . .
                                        Heart then trips a beat
Peering into unguessed taxonomies:
Kekulé, dreaming of the benzene ring,
Let fantasy tell reason where to stroll.

Reason, the dream of dreaming centuries,
Listened to what prodigious Sigmund said,
Dwindling back to a tense meniscus: dread
And traumata beat against that skin,
Language papers vainly over the top—
Try it again with a Not in it, you said.

THE INHERITANCE

Dunked into life, a squalling brat
apeing the role of perfect child,
I let this language buoy me up,
shock troops lightly graduated:
nasty, nice, nectarine, nasturtium, noun.

The stuff was rich as mother's milk.
I couldn't see it didn't fit,
making it do so anyway,
eliding what was grossly wrong.
Origins prove nothing, said William James.

I romped round discourse in my room
only devouring foreign books—
northern, that is—containing heath,
lorries, wolves, bobbies and snow.
The signified was quite inadequate,

a mere Australia. City fathers
had long conspired with Empirespeak
by cancelling native foliage;
so every winter English buds
flashed into fluffy pastel bloom again.

As cunning as a leaning dunny,
this international currency
parades its virtue in old rhymes,
Tomb after tomb, as death with breath.
We swim along with it. We swim and drown.

MENTAL EVENTS

The brain,
a fat grey flower,
blooms
    on a stem
        of bone.

The colours it desires
are unbounded
tropical
    tasteless
        fauve.

It feeds
on darkness
like a flock
    of vampires
        sucking away.

The whitecollar bosses
require it
to shake hands
    with a terminal.

'Up you,'
it cries,
flouncing away
    on clouds
        of homespun gold.

Bouffant, suave, grouse,
orchestral music
by Handel
    skirls through
        the tossing hair.

RIVER RUN

for Kevin Hart

In the beginning was John.
He had shares in the logos
and could treat all that stuff,
antimatter and magic,
with elastic familiarity.

But the name of the legion was legion
so that stocks grew bearish,
world swam under thick waves
of coarsening history,
and the pi-meson said nothing.

A small ache behind my clavicle
may be arthritis,
a bend in the third world
unpick itself as torture,
the inexpressible word.

We puff along like beetroot,
a sweat of scullers following
their coach on his bicycle.
Can I parse the figures of eternity
with my shoulders between my knees?

THE EVOLUTION OF TEARS

I wonder whether grief was already invented
as far back as the Mindel and Würm glaciations,
some rogue gene having tipped the scale.
                                  I wonder
whether the human creature was chosen by grief
as much as by stone tools or opposable thumb.
Did our hopeless remote forebear sit blubbing his eyes out
(her eyes?) by gunyah or cave? We cannot know.
Tears leave no grooves in archaeological sites,
a broken heart has never been trowelled up;
the *lacrimae rerum* do not resemble objects
though misery be as hard as a stone in your hand.

A long way down the line were those other properties,
the lily of logic with a rose between her teeth,
rationalization, debt, the barque of state,
but darker human traits had crept in first
teaching the dread monosyllables, lost and gone.
One, more obscene, begins with a capital D.

Chemical? We comprehend grief, but not always.
A burial rips the guts out of everyone close.
We fail. We suffer. One slides into the ground
who had been a spirit, horsed round and laughed as we do
and now joins the majority . . .
                            the tacit ones . . .
in a horizontal kingdom underfoot.
We do not understand.
                  Together we rise,
feather, turn and fly away.

THE SOUND

'Hush,' you murmured, 'listen.'
And we cocked our heads in a burglar-watching way
Until you decided the coast was clear.
'It's a sound like little chickens hatching,'
You then concluded, made your way upstairs.
I looked out the back window, through the fuchsia,
Toward the overbright stars
And the whole sky was full of chickens falling,
Plump as partridges, gently bobbing down
Or floating over the garage like balloons.

Taking it very much as it comes
Or with a pinch of salt,
Such a fluffy epiphany of chooks
Must have meant something particular after all.

It was a sound like little chickens hatching.

FOR CRYING OUT LOUD

Here
quite as much as there
in the dead straight street
or snailed and breathing gardenplot

time,
that sarcastic medium,
ran silvery through my fingers
like sand, or bonemarrow.

It
leaches through every life
which steps gingerly into it
under and over again.

We
were set down on moist earth
as though to train for some Grand Final
which is never going to take place.

Such
living as you greenly had
flowered and fruited boldly after all
but you misread what it meant.

Now
there's a torrent in the blood,
a sense of arabesques fanning out
across their shining enormous mudbrown delta.

In
the sepia vision, daily, diagonally,
you are walking in mufti backward
through yourself.

PUCK DISEMBARKS

That sun is glazing and glaring from the wrong direction.
In his government regulation gear
And cultural arsy-turvitude
Puck steps ashore in a grammar of ti-tree.
      He rocks the pinnace.
      The foliage looks pretty crook.

Even a spirit can fail to be gruntled
Standing on his northern hemisphere head
In a wilderness without fairies or dairies,
Whose Dreaming he cannot read.
      He tweaks a tar's pigtail.
      This land is all wombat-shit.

The mosquitoes lead him to think of swallows,
The dipping swallows of Devon
And these alien magpies can sing like Titania
In love with a kangaroo.
      Puck waters the gin,
      Peddling the balance to snubnosed natives.

The glittering wavelets throw on yellow sand
Big shells like Wedgwood ware
As the imp rises inside him, getting ready
To rewrite Empire as larrikin culture.
      He daubs a first graffito on
      The commissary tent, GEORGE THE TURD.

Against the pale enamel sky
Rebel cockatoos are screaming
New versions of pleasure:
This is the paradise of Schadenfreude.
      He begins to adore
      The willy wagtail's flirting pirouettes.

THE BUSH

for Seamus Heaney

Overture:
                violins:
it is all scraggy,
wideawake,
                ironical,
decked out
                in denim fatigues.
Witty and welcoming,
                leathery-evergreen,
bemedalled with beercans,
cowpat and wallaby-dung,
flap,
    nub,
        hinge,
            node,
blindeye quartzite,
                wafery sandstone,
bright as a button
subtle for mile on mile
far from vulgarity
                (far from sleek Europe)
in its array of
                furniture tonings
sheeted by sunglaze
                lovingly dusted,
wispy and splintery,
tussocky,
                corduroy,
all of its idiom
dry as a thesis
to moist outsiders:
wonderfully eloquent
                on its home ground,
branchful of adverbs,
lovingly
                wombat-hued,

dreamily
                    sheeptoned,
fluted with scalloping surf
and every step a quip.

AND THE WORLD WAS CALM

Sandbags of sugar cannot conceal the gloomy fact
That we are inserted headlong into life
As a new pen is dipped in lavender ink.
We take up a space amid the comings and goings
Haphazardly, wanly. Velvet wrappings of eternal night
Contain our small blink, pitiless in their Logos.
Powderblue through gentle distance, lyrical mountains
Look at our passing span with incomprehension:
We never read their huge minds, and we die.

Why was the serpent given access to language and stuff?
Awareness becomes a different kettle of fish when eked
Out in a long line: the clipped image gives over
Until your modulation from rambunctiousness to grief
Is felt as a matter of slow brown flow, all river
And no cute islands. Remember, grace yields to Valium
Down on this late-in-the-century flood plain
Where even the fodder and grain crops are postmodern;
That is to say, containing no vitamins.

I'd like to build a cabin or humpy for awareness,
Something rustic like the Bothie of Tober-na-Vuolich
Where I could hum and tootle against the wind,
That long grey stranger skirling through everything.
Grammar is always complete, but the world is not,
Shrugging and folding, surely hatching out of magma.
'You see what you are dreaming, but not with your eyes',
Said a chap who met little sailors down in the park
But grew austere as the case went on wearing him down.

Light is more mysterious than anything else—
That is, except for having to shit and for loving,
Categories you could not have invented, supposing
You were God for a while.
                              Am I alive or dead?
A question basalt or sandstone could never ask.
Awareness becomes a different kettle of herrings
When it's applied to the psyche of a whole country,
Something quite rosy, hydra-headed and fat:
Public opinion did a jig on the carpet of madness.

In the beginning green verbs went bobbing in space
Which was pearly or golden in its painterly turn
And we do not think about gales in the Garden of Eden
Nor about any distinction between plants and weeds
So that Adam is constantly doing something with roses.
Rubric, baldric, erotic, I brood on these terms,
He could have reflected, leaving his pruning aside
While he rolled the well-made words on his tongue like stone:
For the main thing then was learning how to think crudely.

Subtle as ivory handles, we think we are now
Umpteen years on from that scene in the Olduvai Gorge,
Outside the gates of which St Michael struts with his sword.
Pining, we find little paragraphs in which to lament
That we are inserted headlong into life;
Poetry survives with its coppery glint of gnosis
Along one edge.
                    It is a drug that endures
Riding atop the bubbles of evanescence.
The river we step in will burn us off at the knees.

LIKE VIBRATIONS OF A BELL

Very beautiful are the fields of language,
    such happy paddocks,
inkily blushing under those marvellous
pink, barred clouds and angled rays
    we have felt today,
eve of the feast of St Ignatius Loyola,
    the day on which,
as it happens, lovers were struck by lightning.

But that's the way of it, I'm afraid,
    and, speaking of fear,
an old friend lies tonight in St V's,
soon to be torn clean away by cancer,
    the word we cannot say:
it has replaced the devil in our world
    or a black hole
where all Pandora's horrors have been stowed.

Silliness helps to ease the pain, at least
    for those of us
who don't have to clench it in our guts.
Turning to yesterday's *Age*, I can chuckle
    at a couple
married on a shipwreck underwater
    where the guests
slowly showered them with sardine confetti.

All the way from ecstatic to tormented
    my Shorter Oxford
takes its dominant unfeeling way:
it was not born for death. But our dying
    cannot find the words
for someone who shudders near the brink now.
    Come, sweet language,
see him gently under the low hard lintel.

TRACE ELEMENTS

. . . but surely the dead must walk again.
They stroll most oddly in and out of
small corners of your being, optical blips.
They go with an awkward gait, like foreign changelings
through the edges of a crowd or down the block.

It is at random seasons when the mind
is full at ease that my father, roundshouldered,
shuffles along to wait for lights to change
or my tall son shambles down the footpath
in a woollen cap, relentlessly unfashionable
and quiet as a cloud.
                What do they want?
Can they be translated?

Space-time is no longer their medium;
                      they inhabit
antipodes of the radiant fair dinkum,
post-Heisenberg, transphysical, post-Planck,
taunting us all with quips of antimatter.
They are black holes punched in the modern world.
They have been resurrection.
                  They are Dreaming
and we the dream they paint their names across
in grey and lavender and thunderblue,
photocopies of Krishna passing
by Lasseter's reef; or somebody
behind us on the back road to Emmaus,
footsteps in the dust.
                I would not have it
any other way.
        They walk on by.

AFTERNOON IN THE CENTRAL
NERVOUS SYSTEM

Eating raw cabbage at a paper-
littered table at autumn's end
I choose (or something chooses me) to read
an article about biology unclearly,
following in particular the Lamarckian
bit, the easy assault on Skinner
(everybody's enemy, I hope), the
anecdotes about Gregor Mendel's peas
and the delicate paths being traced
through the evergreen mind-body problem,
tracks over tricky terrain as dodgy
as those by which molecular adjustments
carry something from the senses clearly back
to what I, of necessity, call Us.
We, whatever we are, keep wanting
to know how suppositious self in fact achieves
the confidence to keep slogging on
despite the madly random death of cells
and the rupture of connections.
Eating the noisy taste of cabbage while
a CD plays Debussy's *Iberia*
and ash leaves fall on the concrete slabs
of our backyard, I am bemused by how
the musing of the world thus chose me here
out of, say, Scottish tribes and the plaited rush
of history from Plato down to NATO.
Why omnivorous? Why darkish-skinned?
And whence this quaint obsession with ball-games,
as well as making verses? The dumb gene
says nothing at all, but sits at home in my soul
writing me still across its illiterate plan:
a singular man chewing some general cabbage,
looking out across the second millennium
and feeling as fit as a trout.

OPERA WITH PHANTOMS

Demystifiers . . . let it be . . .
We dream because we are lonely asleep,
characters lazily drawn toward us

broken off the stick of childhood
and proffered in loping narrative
which is both saturated and boundless.

Metamorphic as rainbow balls
they dawdle behind us in long rooms
with bell, mirror, postmen, dinosaurs

moving through soft lunar dust,
dipping gently but not too deep
in the residues of yesterday

from bung committee to lethargic tram,
bills, dialectic, loonies, dyslexia
and kids' gear all over the floor.

The dreamer under a low ceiling
works away like a Wonthaggi miner
getting the stuff out:
                                        the lumpy quota.

So there we are, as my aunt would say
were she alive instead of dead,
who had me dream of cyanide for years

by reading from John Dickson Carr
one chapter incorporating
the shadowy primal horror.

FREE WILL

Your choices are dangling in the wardrobe.
They are lavender, navy or grey
as a turtledove rasping midsummer-long
by the cobbled back lane.

Your decision will haunt you,
so fraught with categories
and a thicket of repercussions:
you are in a blue funk.

What do mere shirts have to say
in a wooden world?
They feed into the shady way today
will ravel and unravel you.

A voice that is not of human origin
barracks you down to hell
or through heaven's indigo funnels of
    sheer depth.
It speaks you clean out of mind.

'I am,' you croak (like God himself,
without the horsepower)
but whiplash winds are dashing you now
through undreamed galaxies.

The day slides away and you fall
with a cotton torrent of laundry
into the turquoise-lacquered
Too Hard Basket.

Fight back, feathery faintheart.
It is high time to unfurl again
the enormous rippling spinnaker
of hope.

CALL ME JACK

Toodle-oo, nation-states,
fenced in your vulgar languages.
I am sick to death of you now
like any footling marxist grimly nursing
dreams of a blue-chip world-order.
Pack up your airlines and consuls.
Turn back the enormous clock.

Let us try to picture the Goths
and, lovingly, the Celts
with their soft plaid bags of sentiment.
Look at those little flags
toddled across Europe and its swamps.

Gawking down, the harvest god, Cernunnos,
watches over the milky Celts
crooning in their barleyfields
a little the worse for wear,
giving misdirections to Roman centurions:
'Third on the left past the dolmen.'
Sly wheedlers, they specialize in
druids, the harp, and bronzeware.

Hey, here come the Goths
like platoons of private-school bullies
trudging into the valley, bluffly enquiring,
'Hey, miss, where is your leader?'
Come evening, the musclebound warriors settle down
letting women like Marlene Dietrich
comb butter into their hair.

All over the fringes of empire,
rebarbative as orang-utans,
some gang will come out of the forest
and a stammering ploughman will ask
the chap with the fancy cloak across his shoulders,
'Excuse me, sir, are you a Roman,
a Goth, or a marvellous Celt?'

Then the count slides his runic blade
back in his flash new scabbard
mumbling, 'Just call me Jack'

(or Sextus, or Caradoc).

GARTH McLEOD BROODING

'We are, I suppose, about 30% wrong,'
        the bloke in my daydream said,
dozed over paper piles on a mucky desk.
Consciousness turns around and snaps me back
to the next sub-committee concerning committees
        and the living dead.

Nescafe, a Marie biscuit, the mail
        which never turns out nice
but is merely expressive of people making demands,
confirming my bitter view of human nature . . .
the hopes for human improvement held when young
        will not come twice.

The fax and the xerox are humming a duet
        in their paperhearted yearning,
I feel as old as a beanbag or a Beatle,
some force having emptied me out of my life,
dumping me here to dry, in rank spite
        of all my scheming.

We have to make a submission to the Commission
        by the fourteenth of June.
Did Louis Quatorze have to live like this?
Did Sir Keith Murdoch? Or Sarah Bernhardt?
I could be a dog shut in a small backyard
        howling for the moon.

Our marketing policy feels the hammer of God.
     Our legs in Perth
are walking unknowing into a golden handshake,
having dropped the bundle like a hot potato.
And the Tassie account has gone right down the gurgler,
     no cause for mirth.

Over my Flat-Out Basket the golden motes
     are in fine array,
my cufflink clunks; collar's a whisker tight;
I am not really what I am,
but a lumpish teenager whose contemporaries
     have suddenly turned grey.

Slowfoxing at Power House when I was eighteen
     I fancied my zest
would maintain its voltage more or less for ever
but every optimist is a dunderhead . . .
We must get minutes out to Jenkins and Diorno;
     right now is best.

A baffled blowfly bats against the glass.
     Soft elms beyond
are wearing a snood of sunlight on their hair
while I niggle at the budget. Two weeks hence
we'll be struggling through an Annual General Meeting
     like a mud-choked pond.

SUNSET SKY NEAR COOBER PEDY

Streak, dash, fluff, apricot radiation,
the whorled blurring of a burned edge:
solid lavender continents and filiations,
islands in air, foolishly soft fleeces
dandled overhead, easterly pinks and peaches
nervously barred with mauve.
                       Dollop, smear,
streaking and massing, pos and neg,
versions of Crete, New Britain and the Coorong,
wool, flax, gelato, soapy froth
teasingly spread around like transient solids
over, behind, permeating or through.

They are done by drunken painters of genius
who, visited by vast hallucinations,
daubed them all over a monster's mural hall.
Some are like dried-out corpses glorified
or windblown spare parts of the heavenly host.
Their grammar escapes me.
                      The tune is hard to hear.
Powderblue, cream, blush, incandescent copper,
the meaning of what they are is merely IS.

LOOKING DOWN ON CAMBODIA

Arabesque 3D shoreline
against an overwhelm of morning haze
with a silver delta threading its liquid gush
beyond those matted islands,
like reptiles underneath our Boeing home.

You can't see the blood;
you can't hear the dying
or mounded skulls rubbing faintly together.

Sea is like Thai silk
swathing life's ragged margin
so that the peace
of microscopic praus
and signatures of islands throbbing by
obscure ferocious fact:
they are duck and beanshoots in a long steel sea.

The Khmer Rouge killers
are stabbing still,
down somewhere out to my far left.
If only some great thumb could rub them out
of mind, for ever.
If only history could take another fork,

a silver fork
prodding the tranquil sea.

REALITY

A lost boy in another body
I reach yearningly backward, Dad,
keen to capture you entire,
if memory permits entirety.

Now from the oarbench of 'La Mouette',
bobbing lightly, I see the real you,

fifty-odd, fit, portly, busily leaning
over the stubborn outboard motor
tugging the starter-cord.

It coughs.
        No. Mt Martha rests
steady on the shallow amber beyond.
You jerk the cord again.

The little motor roars. Throttle turned back,
our dinghy takes off with you
swinging it round
and peering over my left shoulder
westerly to sea
from under your tennis eyeshade.

Out across the inshore sandbanks,
warm in faintly rumpled water,
a creelful of flathead loaf,

dreaming of capture.

You turn the motor off,
breaking the brittle whitebait slowly
but the dumb tides of memory
let you go.

DUSKY TRACTS

It was the coarsening edge
of the dream attacked me,
thickened with awareness
of enormous dark.

It was the pinegreen melody
that bears no language
began to hurt my knowing,
hard as a wall.

It was blotchy glimpses
of an inkblue deep
as I bobbed all alone
over wavewracked fathoms.

Nothing got said
in that salt orchestration:
strings of the cello
were drawn out of my guts.

Withdrawn from the story-line
was He Who Is,
every image bespeaking raggedly
that which is not.

ODE TO MORPHEUS

It's a rum go, a pretty pickle, a rare kettle of fish
    that we spill so much of our time
(I will not say our days, those branchy olive intervals)
    rocking away in your arms: pointblank there.
Well, not as blank as that, but numbly a-wander
    along your tracks, your labyrinthine
halls through the great, ivied house, that is
    always oneself in the long haul
studded with metamorphosis and yellow anticlimax—
    the wide, free mini-series of the night.

It seems pretty weird, oddball, queer as a coot
    that we switch off a third of our days
swaddled in linen, squeezed between counterpane
    and kapok; perhaps under doonas.
There was a bloke in the Odd Spot who somehow got by
    on 1.5 hours per night.
Was he smarter than us? We'll never suss it out
    but I resent like crazy all of this
rehearsing for peace-without-end in our long last home.

It's a hard god, a crook umpy, a two of spades,
    that figured our fortune out this way.
Instead of ranging the night, we snooze in your lap,
    the years ticking away like clockwork ducks
or hurdling the fence like sheep.
                  Fiddling the hints . . .
    sheedling the woollen flump . . .
                       then, zzzz . . .

MORPHEUS REPLIES

Come into my arms, you corpse-like infant:
I'll tell you such tales, they will blow your mind.

Having lain down softly, worried only by
adjusted position of elbow and knee, by

tucking in of clean sheet, its turning down,
you have been given over to me. Stet.

Lie there in your stupendously dreamed
blanket of innocence. Snooze like a monster

or some prodigious rock, sucked by bottlegreen waves.
This will be your disco of horrors and blisses.

You are mine, stupid. Even wise, I have you:
the night will pour you full of blazing lamps,

will suck 200 volts along your bonemarrow
and make old heart bark like a dreaming dog.

Take it easy; regard kapok as the world
and float.　　　This is my archipelago.

WHAT ARE THESE COMING TO THE SACRIFICE?

I thought it all so strange,
this living we went on with,
full moons wheeling past in their turn
and the heady jasmine times.

One day we were all young,
a tennis ball thrown on the beach
and bare feet ruffling
edges of insolent waves,

and then quite ungainly
creatures of middle ageing,
grumpy and stiff as we bent over
to pull on shoes and socks.

Very few ever caught a glimpse
through the ontological window,
its dusty glass reflecting bills
and a coffee cup.

Somebody raised his head
from all those yellowing papers,
in time to see a brilliant rosella
flare through the wet grevilleas

as life still might do
quite unwittingly.
                    We are
vulnerable to bruising
like perfect raspberries.

WHY DO WE EXIST?

The child sits, quiet as a moth,
under murmuring trees in the garden,
a blackbird warbling grandly,
wrens and wattlebirds
doing their various things
overhead and around,

and the child knows
he is very small in the garden,
smaller still in the world,
as nothing in the . . .
how do you call it? . . .
       universe.
So that his being there,

fragile in a rustling suburban garden
among heaving ripples of green,
is a kind of miracle.

In the end he is grateful.

OXFORD POETS

Fleur Adcock
Moniza Alvi
Kamau Brathwaite
Joseph Brodsky
Basil Bunting
Daniela Crăsnaru
W. H. Davies
Michael Donaghy
Keith Douglas
D. J. Enright
Roy Fisher
Ida Affleck Graves
Ivor Gurney
David Harsent
Gwen Harwood
Anthony Hecht
Zbigniew Herbert
Thomas Kinsella
Brad Leithauser
Derek Mahon

Jamie McKendrick
Sean O'Brien
Peter Porter
Craig Raine
Zsuzsa Rakovszky
Henry Reed
Christopher Reid
Stephen Romer
Carole Satyamurti
Peter Scupham
Jo Shapcott
Penelope Shuttle
Anne Stevenson
George Szirtes
Grete Tartler
Edward Thomas
Charles Tomlinson
Marina Tsvetaeva
Chris Wallace-Crabbe
Hugo Williams